Education Through Cooperative Extension

Education Through Cooperative Extension

Brenda Seevers
Donna Graham
Julia Gamon
Nikki Conklin

Delmar Publishers

an International Thomson Publishing company I(T)P®

Albany • Bonn • Boston • Cincinnati • Detroit • London • Madrid
Melbourne • Mexico City • New York • Pacific Grove • Paris • San Francisco
Singapore • Tokyo • Toronto • Washington

NOTICE TO THE READER

Publisher does not warrant or guarantee any of the products described herein or perform any independent analysis in connection with any of the product information contained herein. Publisher does not assume, and expressly disclaims, any obligation to obtain and include information other than that provided to it by the manufacturer.

The reader is expressly warned to consider and adopt all safety precautions that might be indicated by the activities herein and to avoid all potential hazards. By following the instructions contained herein, the reader willingly assumes all risks in connection with such instructions.

The publisher makes no representation or warranties of any kind, including but not limited to, the warranties of fitness for particular purpose or merchantability, nor are any such representations implied with respect to the material set forth herein, and the publisher takes no responsibility with respect to such material. The publisher shall not be liable for any special, consequential, or exemplary damages resulting, in whole or part, from the readers' use of, or reliance upon, this material.

Cover art courtesy of : Digital Stock
Cover Design: McKinley Griffen Design

Delmar Staff
Publisher: Tim O'Leary
Acquisitions Editor: Cathy L. Esperti
Senior Project Editor: Andrea Edwards Myers
Production Manager: Wendy A. Troeger
Marketing Manager: Maura Theriault

Printed in the United States of America

For more information, contact:

Delmar Publishers
3 Columbia Circle, Box 15015
Albany, New York 12212-5015

International Thomson Publishing—Europe
Berkshire House 168-173
High Holborn
London WC1V 7AA
England

Thomas Nelson—Australia
102 Dodds Street
South Melbourne, 3205
Victoria, Australia

Nelson Canada
1120 Birchmount Road
Scarborough, Ontario
Canada M1K 5G4

International Thomson Editores
Campos Eliseos 385, Piso 7
Col Polanco
11560 Mexico D F Mexico

International Thomson Publishing—GmbH
Königswinterer Strasse 418
53227 Bonn
Germany

International Thomson Publishing—Asia
221 Henderson Road #05-10
Henderson Building
Singapore 0315

International Thomson Publishing—Japan
Hirakawacho Kyowa Building, 3F
2-2-1 Hirakawacho
Chiyoda-ku, 102 Tokyo
Japan

1 2 3 4 5 6 7 8 9 10 XXX 02 01 00 99 98 97

Library of Congress Cataloging-in-Publication Data

Education through cooperative extension / Brenda Seevers . . . [et al.}.
p. cm.
Includes bibliographical references and index.
ISBN 0-8273-7172-1
1. United States. Extension Service. 2. Agricultural extension work- -United States. I. Seevers, Brenda.
S544.E48 1997
630' .71'5073--dc20

96-23113
CIP

Contents

Preface ix
Acknowledgments xi
About the Authors xiii

Chapter 1 **The Scope of Cooperative Extension: Mission and Philosophy** 1
Key Concepts 1
The Cooperative Extension System 1
Philosophy 5
Mission 7
Summary 11
Discussion Questions 11
References 12

Chapter 2 **The Origin of Extension Work** 15
Key Concepts 15
Agriculture: The Nation's Central Pursuit 15
A College for the Common People 19
The Idea of Extension Education 25
The Smith-Lever Act 35
Summary 40
Discussion Questions 42
References 42

Chapter 3 **Organization, Structure, and Administration for Extension Programs** 45
Key Concepts 45
Organization and Structure 45
Staffing in Extension 50
Employment in the Cooperative Extension Service 53
Training and Orientation 54
Assessment of Performance 58
Legal Aspects of Employment and Programming 59
Administration of Extension Programs 61
Management, Leadership, and Supervision 62
Summary 66
Discussion Questions 66
References 67

Chapter 4 **Program Areas in Cooperative Extension** 69
Key Concepts 69
Traditional Program Areas 69
Agriculture 70
Home Economics 75
4-H/Youth Program 78
Community Development 85

Interdisciplinary Programs and Other Program Areas....87
Summary....88
Discussion Questions....89
References....89

Chapter 5 **Developing Extension Programs....91**
Key Concepts....91
What is a Program?....91
Key People in the Program-Planning Process....92
Factors Influencing Program Development....94
Exploring the Steps in Program Planning....97
Program Design and Implementation....106
Program Evaluation....108
Involving People in Program Development....109
Program Management Strategies....113
Communicating Program Plans....114
New Directions in Extension Program Planning: Issues Programming....114
Ethical Issues in Program Planning....116
Programming Pitfalls....118
Summary....119
Discussion Questions....119
References....119

Chapter 6 **The Teaching-Learning Process....121**
Key Concepts....121
The Concept of Education....121
The Nature of Adult Education....123
The Teaching-Learning Interaction....124
The Learning Process....132
Why People Participate in Education....134
Planning for Success....136
Summary....136
Discussion Questions....137
References....137

Chapter 7 **Extension Teaching Methods....139**
Key Concepts....139
Useful and Practical Instruction....139
Extension Methods....141
Individual Contact Teaching Methods....143
Group Methods....148
Mass-Media Methods....155
Selection of Teaching Methods—Technological Impact....161
Summary....163
Discussion Questions....163
References....163

Chapter 8 **Evaluating Extension Programs** **165**
Key Concepts 165
Introduction 165
Reasons to Evaluate 167
Managing an Evaluation 168
Steps in Evaluation 170
Objectives-Based Evaluations 171
Data Collections Methods 175
Group Techniques for Needs Assessment and Evaluation 176
Analyzing the Data 182
Reporting the Results 184
Summary 184
Discussion Questions 185
References 185

Chapter 9 **Management of Volunteer Programs** **187**
Key Concepts 187
Volunteers in Extension 187
Volunteer Program Management 190
Developing a Volunteer Management Program 192
Future Trends and Issues 199
Summary 200
Discussion Questions 201
References 201

Chapter 10 **International Extension** **203**
Key Concepts 203
Introduction 203
Reasons to Study Extension in Other Countries 203
Origins of Extension Worldwide 205
Comparison of U.S. Extension and Extension in Other Countries 206
Donor Agencies 208
Extension Approaches in Developing Countries 209
The Chinese Model 213
Extension in Western Europe and Canada 215
Eastern Bloc and Former USSR Countries 216
Problems and Constraints 217
Summary 221
Discussion Questions 221
References 222

Chapter 11 **Extension Future Focus** **225**
Key Concepts 225
Extension in a Complex World 225
Societal Trends 226
Assessments of Extension 228
Future Focus 232

The Challenge Ahead 237
Summary 238
Discussion Questions 239
References 239

Glossary 243

Appendices

Appendix 1: Smith-Lever Act (May 8, 1914) 253
Appendix 2: Smith Lever Act (Amended November 28, 1990) 257
Appendix 3: Chronological History of Legislation by the United States Congress Relating to the Cooperative Extension Service in the Land-Grant Universities 263
Appendix 4: A Profile of Extension's Professional Staff (1914–1994) 271
Appendix 5: The 105 Land-Grant Colleges and Universities 273
Appendix 6: U.S. Department of Agriculture Headquarters Organization 277

Index 279

Preface

Cooperative Extension is the world's largest nonformal educational organization and is widely recognized for its success in addressing the concerns of a changing society. This book provides a comprehensive overview of the organization, operations, and programs of Cooperative Extension. Although the primary emphasis is on current educational approaches, we've also provided a thorough portrait of Extension's foundations, history, and philosophy. Although designed as an introductory text for undergraduate classes on nonformal education, we expect it to be useful for a wide range of audiences. The book should be ideally suited as a preservice text for those interested in pursuing a career in Cooperative Extension. Administration may wish to provide copies as part of the orientation and training for new Extension employees. Staff in county and parish offices should find it useful in working with councils and committees who are unfamiliar with Extension's organization and structure. Another potential audience consists of international students and dignitaries who wish an overview of the U.S. Cooperative Extension System. This text will help explain how Extension operates. And finally, some of us expect to use it in our graduate classes for students to critique and revise.

Each of the authors has extensive experience with Cooperative Extension. Our starting dates for careers in Extension go back as far as mid-century and continue to this day. Our experiences include county, area, and state positions, and all of the program areas. All of us teach courses related to Extension and stay current with new approaches to nonformal education. We've put together a book that we can use ourselves and are excited about sharing it with others.

Chapter overview:

Chapter 1. A brief introduction to the organization of Extension plus a discussion about its philosophy and mission.

Chapter 2. A history of factors that influenced the development of Extension as well as how it has evolved into the organization we know today. A number of historical photographs add interest.

Chapter 3. An in-depth view of Extension's current organizational structure and administration. Discussion includes common funding and staffing patterns and the management process.

Chapter 4. A description of the major program areas (agriculture, home economics, 4-H/Youth, and community development) that undergird most activities and educational efforts.

Chapter 5. The program planning process; a review of several models and steps. This chapter and the next three chapters provide details on how to develop, implement, and evaluate nonformal educational programs.

Chapter 6. Factors in the teaching and learning process. Educational theories are related to practices applicable for nonformal learning situations.

Chapter 7. Details on the wide variety of delivery methods used in Extension, including new technology.

Chapter 8. Evaluation of programs, measuring program effectiveness and accountability.

Chapter 9. Volunteers as a vital component of Extension education, with pointers on managing an effective volunteer program.

Chapter 10. An overview of the different structures of Extension internationally. We are living in a global society, and many countries have developed Extension programs to disseminate information, particularly on agriculture.

Chapter 11. Societal and internal forces influencing the direction of Extension. Although none of us can predict the future, we have investigated the forces that have influenced recent changes and those that might cause changes in the future.

Cooperative Extension is a unique and complex entity. We have attempted to provide a current assessment of organization and methods; it is as in-depth as possible for the body of one text. We have appreciated the help of reviewers, students, and colleagues, but take full responsibility for any mistakes or omissions. Hopefully, this text has accurately described the truly one-of-a-kind institution that is Cooperative Extension.

—The Authors

Acknowledgments

The development of a book requires the assistance of many. This book was no exception. The authors would like to express their appreciation to the many individuals involved in completing this text.

Special thanks goes to the following people who provided technical assistance and information:

Dr. Paul D. Warner, Associate Director, Kentucky Cooperative Extension Service, provided the most current reports on the structure and function of the Cooperative Extension Service.

R. Warren Flood, Curriculum Materials Service, Department of Agricultural Education, The Ohio State University for photographic design, and

Lisa Southern and Jim Keene, University of Arkansas, for their secretarial assistance in typing and editing the manuscript.

In addition, appreciation is expressed to those who provided suggestions and/or reviewed part or all of this text. Their suggestions and comments were much appreciated, but they should not be held accountable for any omission or errors.

Lisa Lucas, University of Arkansas

Cathy Montes, Assistant Editor, Agricultural Communications, New Mexico State University

Jimmy G. Richardson, Ph.D., and his students at Mississippi State University

Mary Foley, Extension Sociologist, Iowa State University

Larry Trede, Ph.D., Associate Professor, Department of Agricultural Education & Studies, Iowa State University

B. Lynn Jones, Ph.D., Associate Professor, Extension Evaluation, Iowa State University

Yong Hwan Lee, Ph.D., Professor, Seoul National University, Korea

Jane Ann Stout, Program Leader, Extension to Families, Iowa State University

Joseph Kurth, Ph.D., Program Leader, Extension to Youth and 4-H, Iowa State University

Paul Coates, Ph.D., Program Leader, Extension to Communities, Iowa State University

Bill Edmundson, Economic Research Service, Washington, D.C.

Finally, many students have assisted with the writing of this text. Grateful appreciation is due the following people who were students in Julia Gamon's classes at Iowa State University: Gail Grant, Mitch Hoyer, Tim Wilcox, Arnold McClain, Dur Mohamed Abbasi, David Bjorneberg, Dorothy Chestnut, Patrick Derdzinski, Sami El-Ghamrini, T. K. Khatib, Dawn Mellion, Chuck Morris, Gary Wingenbach, Chia-Hsing Wu, Jane Hayes-Johnk, Kristin Blum, Ismail Mohamed, Dawn Hildebrandt, Maria Clark, and Ma'Mun.

Acknowledgments

About the Authors

Dr. Brenda Seevers is an Assistant Professor in the Department of Agricultural and Extension Education at New Mexico State University. She is involved in teaching and advising bachelors and master students in agricultural and Extension methods and philosophies. Research areas have also been related to Extension education. She has worked as a county Extension agent in Saratoga, New York and a state 4-H specialist in Wyoming. She holds home economics degrees from two Ohio schools, Bluffton College and Kent State University, and is a 1991 graduate with a Ph.D. in Agricultural Education from The Ohio State University.

Dr. Donna L. Graham holds a joint appointment as an Associate Professor in the Department of Agricultural and Extension Education at the University of Arkansas-Fayetteville and as Extension Education Specialist with the Cooperative Extension Service. She has had over 20 years of Extension experience at the county, area, and state level. She is involved in seven professional and honorary organizations and has served in numerous leadership roles of the state and national affiliates. Dr. Graham received the Bachelor of Science and Master of Science from the University of Arkansas and the Doctorate of Philosophy from the University of Maryland.

Dr. Julia A. Gamon teaches Extension-related courses and conducts research at Iowa State University, Ames. Her Extension background started with summer internships and began officially with a County Extension Home Economist position. She spent ten years as a county youth agent and supervisor of community development summer aides, four years on the state 4-H staff in Iowa. She is a Professor in the Department of Agricultural Education and Studies, where she teaches both undergraduate and graduate agricultural Extension courses. The main thrust of her research is program evaluation. Six doctoral students and nineteen masters students have conducted Extension-related studies under her direction. She has published 36 articles and given 43 regional and national presentations.

Dr. Nikki Conklin works with The Ohio State University Extension Human Resources Team in Staff Development. She also holds a courtesy appointment in the Department of Agricultural Education, teaching graduate program development classes and working with masters and doctoral students in Extension Education. She has worked in a variety of positions in Extension including county Extension agent, District Supervisor-in-Training, and State Leader for Program Development. She holds three degrees from The Ohio State University.

C H A P T E R

> *"What makes a nation firm and great and wise is to have education percolate all through the people."*
>
> —SEAMAN KNAPP

The Scope of Cooperative Extension: Mission and Philosophy

Key Concepts

- Overview of the Cooperative Extension System
- Profile of the Cooperative Extension System
- Philosophy of Cooperative Extension work
- Mission and scope of the Cooperative Extension Service

THE COOPERATIVE EXTENSION SYSTEM

The Cooperative Extension System (CES) is a public-funded, nonformal, educational system that links the education and research resources of the United States Department of Agriculture (USDA), land-grant universities, and county administrative units. The basic mission of this system is to enable people to improve their lives and communities through learning partnerships that put knowledge to work (Extension Committee on Organization and Policy, 1995).

Cooperative Extension, established by the Smith-Lever Act in 1914, was designed as a partnership of the USDA and the land-grant universities, which had been authorized by the federal Morrill Acts of 1862 and 1890. State legislation enabled local governments in each county to become the third legal partner of this educational endeavor.

Extension is believed to be the world's largest adult and youth out-of-school, nonformal educational organization (Fiske 1989). This educational system includes professionals at fifty-one 1862 land-grant universities, Tuskegee University, and sixteen 1890 land-grant universities. In 1994, twenty-nine tribal-controlled community colleges and higher education institutes that have partial land-grant status under the Equity in Educational Land Grant Status Act of 1994 were added as public institutions involved in this Extension system.

The staff of the Cooperative Extension Service numbers over 16,000, with 64 percent of these professionals located in the 3,154 counties or parishes while the other faculty, normally housed on land-grant campuses, serve as state specialists in various subject-matter disciplines. These professionals work along with thousands of paraprofessional staff and some three million volunteers to assist in extending the educational programs under the oversight and training of Cooperative Extension staff, see figure 1–1.

At the national level, the federal partnership includes an Extension administrator and a group of professionals within the USDA. These professionals direct federal fund allocation, coordinate national initiatives, provide program leadership, and facilitate linkages between the USDA and Congress. This unique federal-state-local partnership has functioned effectively for almost a century as interdependent, yet autonomous in funding,

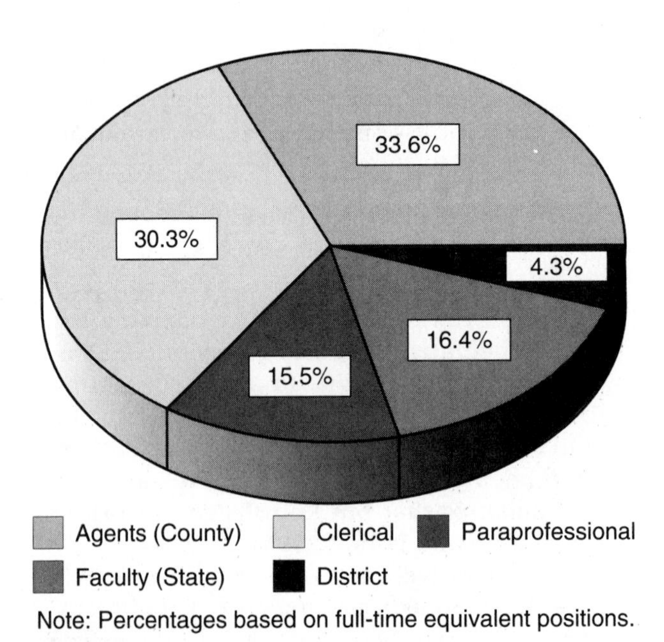

Figure 1–1 ***Classification of Cooperative Extension System Employees. County Office Study of USDA Field Offices (1992). USDA.***

staffing, and programming. Each partner performs distinctive functions essential to the operation of the total system. Cooperative Extension links the research efforts of the USDA and the land-grant universities in order to make scientific knowledge available to all who need it. The system is characterized by communication between those who work for Extension and those who utilize the information—the researcher, Extension educator, and the public. It is a dynamic, ever-changing organization pledged to meeting the country's needs for research, knowledge, and educational programs to enable people to make practical decisions that can improve their lives.

The Name

Initially, the organization was called Agricultural Extension Service but many states have changed their organizational name to Cooperative Extension Service to more accurately reflect the nature and function of the organization. Sanderson (1988) feels this name is most descriptive.

It is *cooperative* that includes three distinct but related and coordinated partners. The federal partner is the Extension Service of the USDA now organized within the Cooperative State Research, Education, and Extension Service (CSREES). The state partner is the Cooperative Extension Service of each state and several U.S. territories. The county or local partner is the city or county government, or other elected authority governing local Extension programs. It is also cooperative in that the partnership is a coordinated effort among three levels of government with three sources of public funding and three levels of perspectives on mission, goals, and priorities for programming, see figure 1–2.

It is an *extension* of the USDA and the land-grant institutions of each state—the outreach partner of the land-grant institution with a role of reaching people and extending knowledge and other resources to those not on campus. Extension programs are offered in homes and communities to address needs, problems, or issues of the local clientele.

It is a *system*, a unique national educational system that draws on the expertise of the federal, state, and local partners to provide practical, unbiased information produced by the research centers and universities to the people. This information, in the form of noncredit education, is provided as a *service* of the university to help people identify and solve problems in their lives.

Profile. The Cooperative Extension Service is unique. No other educational system involves so many levels that are interrelated, yet autonomous. The following characteristics developed by Sanders (1966) and revised by Prawl (1984) provide a profile of the organization and its work:

- Extension is an agency of government created by law.
- It is an agency that provides services to any person without discrimination.

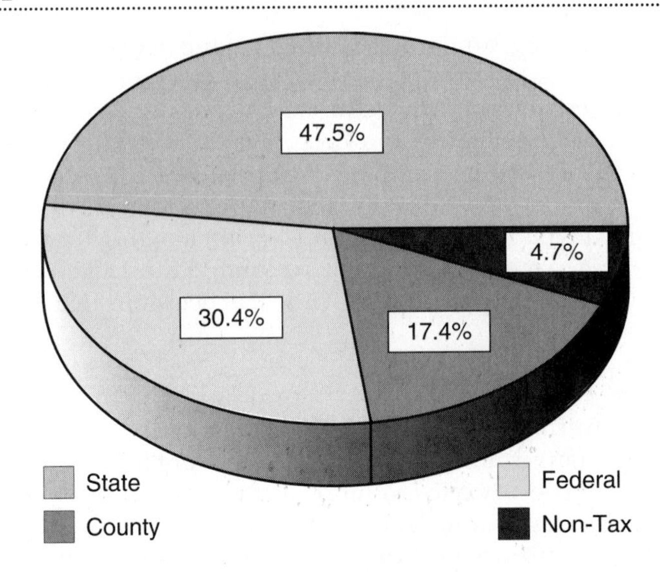

Figure 1–2 ***Source of Funding for Cooperative Extension System. Profile of Extension (1994). Planning, Development, and Evaluation Staff. Extension Service-USDA. Washington, D.C.***

- It is a cooperative with federal, state, and local governments having a part in its administration.
- It is an educational institution that differs greatly from the common mission of an educational institution in that it:

  has no fixed curriculum or course of study,

  confers no degrees and gives no diplomas,

  operates informally off campus and uses farms, homes, and places of business as classrooms,

  uses instructors with a wide range of subject-matter expertise,

  has a large and heterogenous audience,

  offers subject matter that is more practical than theoretical for immediate application to the solution of problems,

  is educational in nature with teaching conducted informally using a wide variety of teaching methods, and

  requires a change of both mental and physical behavior for the application of the subject matter.
- The clientele participation is purely voluntary.
- A technical service is provided to the clientele.
- The field of work is broad and primarily grounded in agriculture, home economics, and related areas.

- The people with whom it works are of supreme importance.
- Programs are based upon needs and expressed desires of the people and communities.
- The information provided is research based.
- It is dedicated to work with the family.
- It is an equal partner with the research and teaching units in the land-grant university system.
- It depends upon volunteer leaders who help plan, implement, and evaluate educational programs.
- Its programs are flexible and valuable in emergencies.

Philosophy

A philosophy is a body of principles underlying a human activity, ordinarily with the implication of practical use, such as a philosophy of life. Four great principles form the philosophical base of Extension work. These are that (1) the individual is supreme in a democracy, (2) the home is the fundamental unit in a civilization, (3) the family is the first training group of the human race, and (4) the foundation of any permanent civilization must rest on the partnership of the people and the land (Bliss 1952, p. 172). These beliefs have helped frame a practical philosophy that has guided Extension professionals, their educational programming efforts, and the practices of Extension work throughout its history. This philosophy is based on the belief that learning put to use leads to a better life for the individual, family, and community.

The philosophy of Extension work was greatly influenced by the tenets of Western civilization—a strong belief in the equality of people, the possibility for change or progress, the reliability of science, and the power of education (Ward 1962). All of these philosophical beliefs were important when the Smith-Lever Act was passed in 1914. In a democracy it is important to maintain as nearly as possible equality of opportunity for all the citizens. People have an inherent or natural right to participate in the benefits derived from what they create and support. When Extension was created, few could afford to go to college and it was thought to be morally obligatory for the government to provide ways of making the practical benefits of the colleges available to the people. The extension of university resources through the outreach system of Extension work is now recognized as one of this country's greatest contributions to democracy (Bliss 1952).

Much of the philosophy of Extension is based upon the well-defined beliefs of James Wilson, the Secretary of Agriculture under President William McKinley, and Seaman Knapp, one of Extension's most influential pioneers. Knapp inspired early agents with a fundamental belief that their value was not in what they could do, but in what they could get other people to do. Knapp and Wilson believed that science and education should be brought as nearly as possible onto every farm, and into every home, and into all the lives of all those engaged in agricultural enterprise. Both believed that those that best served the country were those that best served its rural life. They had a passion for service to the masses of people whose welfare and happiness constituted the stability of democracy. Knapp's work is described in detail in Chapter 2.

Statements of Extension's mission expressed by early Extension pioneers provided the foundation for how these fundamental beliefs were conceptualized and formulated into action. J. Neil Raudabaugh, a federal extension administrator, felt that Extension work was to be based on three principles: reaching people where they are (educational level, interest, and understanding); teaching people to determine their own needs; and teaching people to help themselves (Prawl 1984). Slogans such as "helping people help themselves" and "learning by doing" are examples of this philosophy still in use by the Extension system today.

Extension work was also greatly influenced by the philosophical viewpoints of adult educators such as John Dewey. Various schools of thought advocating different educational teaching and learning philosophies are evident in Extension programming efforts and practice. These philosophies have differing assumptions about human nature, the purpose of education, and the roles of the instructor and learner. Examples of the use of each educational approach used in Extension work include:

Progressive	problem-solving approach
Liberal	content mastery instruction
Behaviorist	systematic instruction models
Humanist	participatory approach of learning
Radical	community change (White and Brockett 1987).

Values and Beliefs

A faith in the individual, the home, and the family formed the common values and beliefs of early Extension leaders. These values are still important today. Sanderson (1988) says that the ultimate value that guides Extension work is the belief in the development of people. Boone (1989) says that the basic belief in the empowerment of people, the importance of rural life, and faith in the future guide the work. Many of these principles are embodied in the "Extension Professionals' Creed" developed by Epsilon Sigma Phi, the professional honorary society for Extension workers, see figure 1–3. Studies by Safrit (1990) and Safrit, Conklin, and Jones (1992) found that integrity, credibility, helping people, excellence, and useful programming were common values of Extension educators. According to a sample of extension employees, values that are framing the future are collaboration, credibility, democracy, diversity, learner-centered, lifelong education, scholarship, self-reliance, and teamwork (Extension Committee on Organization and Policy 1995).

I BELIEVE in people and their hopes, their aspirations, and their faith; in their right to make their own plans and arrive at their own decisions; in their ability and power to enlarge their lives and plan for the happiness of those they love.

I BELIEVE that education, of which Extension is an essential part, is basic in stimulating individual initiative, self-determination, and leadership, that these are the keys to democracy and that people, when given facts they understand, will act not only in their self-interest but also in the interest of society.

I BELIEVE that education is a lifelong process and the greatest university is the home; that my success as a teacher is proportional to those qualities of mind and spirit that give me welcome entrance to the homes of the families that I serve.

I BELIEVE in intellectual freedom to search for and present the truth without bias and with courteous tolerance toward the views of others.

I BELIEVE that Extension is a link between the people and the ever-changing discoveries in the laboratories.

Figure 1–3 Extension Professionals' Creed (1995). Epsilon Sigma Phi.

By understanding the philosophies that have guided Western Civilization and education, one can see how these have had great influence on the mission and scope of Extension work. Each has provided a framework upon which the organization has developed and implemented its educational programming efforts. This is illustrated in figure 1–4.

MISSION

The mission of Extension work, set forth clearly in the Smith-Lever Act of 1914, was ". . . to aid in diffusing among the people of the United States useful and practical information on subjects relating to agriculture and home economics, and to encourage the application of the same . . ." Furthermore, it said that ". . . extension work shall consist of the giving of instruction and practical demonstrations in agriculture and home economics to persons not attending or resident in said colleges in the several communities, and imparting to such persons information on said subjects through field demonstration, publications, and otherwise. . . ." Today, this mission is stated as ". . . the development of practical demonstrations of research knowledge and giving of instruction and practical demonstrations of existing or improved practices or technologies" (Smith-Lever, 1985 amended).

Congress created the Extension Service as a vehicle for human development through nonformal, off-campus education aimed at a rural population. Thus, Extension developed a curriculum based in agricultural sciences, home economics, and youth development as the instrument for meeting the most urgent needs of the time. But Extension's mission and

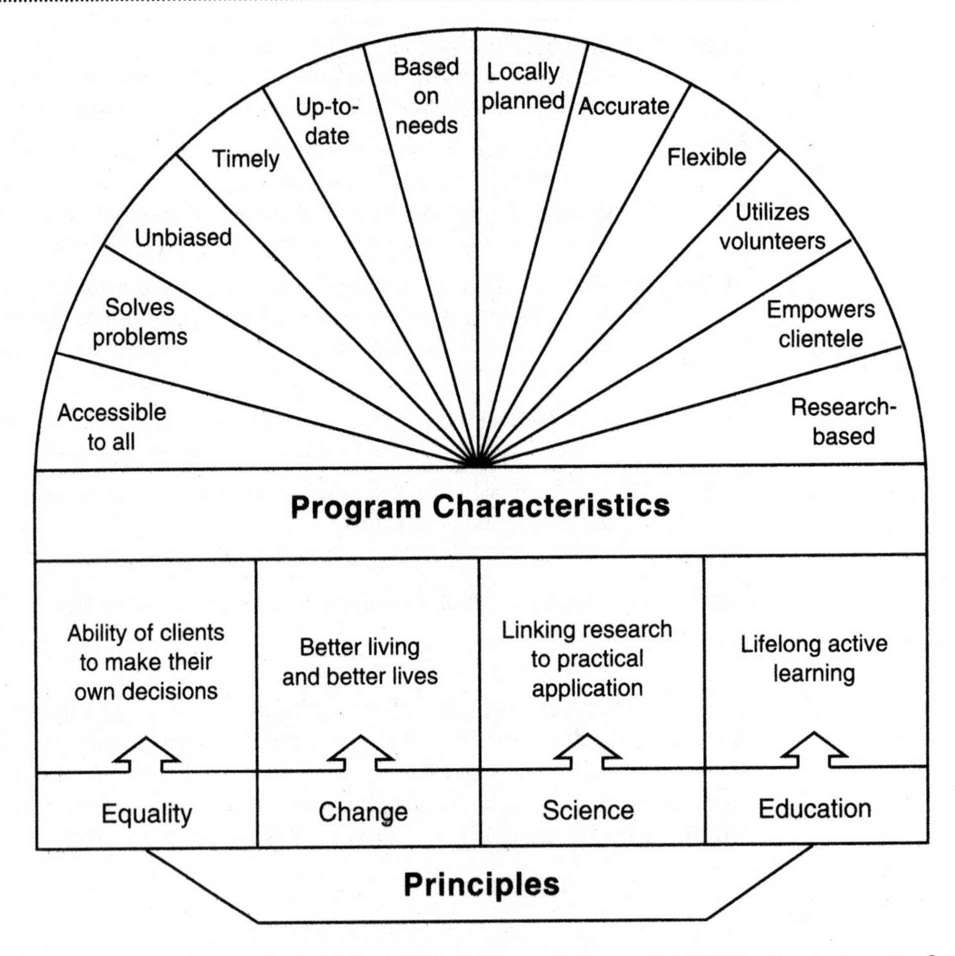

Figure 1–4 ***Extension's Philosophical Base. Gamon, J. (1995). Department of Agricultural Education and Studies, Iowa State University, Ames.***

scope have changed through the years. The *Kepner Report* of 1946, the *Scope Report* of 1958, *A People and a Spirit Report* of 1968, *Extension in the '80s*, and *Framing the Future* (1995) are examples of comprehensive studies regarding the evolving mission of Extension. Societal factors have forced the system to encompass a broader focus and an expanded clientele. The mission of Extension educators has broadened, making them more than instructors in agriculture and home economics. Their role is that of educational missionaries transforming the quality of people's lives and contributing to their development as human beings through education. The current mission of the Cooperative Extension System states that Extension shall enable people to improve their lives and communities through learning partnerships that put knowledge to work (Extension Committee on Organization and Policy, 1995).

At the federal level, Extension is guided by a mission established by the Cooperative States Research, Education, and Extension Service. This mission is one that seeks to advance a global system of research, extension, and higher education in food and agricultural sciences and related environmental and human sciences to benefit people, communities, and nations. (Cooperative States Research Education and Extension Service, 1995).

Some states have a one word mission, such as "education" or "service." Other states use broader mission statements such as "to extend the resources of the university out to the people in the state" and "to help people make decisions" or "to improve the quality of life." Some mission statements are directed toward limited-resource farmers and their families or to specific geographic areas.

Historically, the organizational mission was accomplished through the specific mission, goals, and objectives of four program areas. Since 1987, the mission of the system has focused on national initiatives. It has sought to improve lives through an educational process that uses scientific knowledge focused on issues and needs (Rasmussen 1989). However, base programs are still central to the mission of helping people improve their lives and are administered and delivered through the four traditional program areas—agriculture, home economics, 4-H youth, and community development—within the states.

Agriculture. The mission of today's agricultural and natural resource programs continues to focus on providing an adequate supply of food and fiber for consumption and export by the United States. Public concerns relative to food safety, water quality, conservation of natural resources, environmental risks, ethical applications of biotechnology, and others have heightened the importance of educational emphases in agriculture. Educational objectives have lessened in the area of production agriculture but increased in sustainable and profitable practices. "Society has become much more involved in determining how agriculture will be practiced. Taxpayers and consumers are as interested in the integrity of the environment and the aesthetics of the countryside as they are in assuring the domestic food supply" (Bloome 1992).

Home Economics. The improvement of family and economic well-being, nutrition, and health are major objectives of the home economics program designed for family members. An overall mission is to strengthen the family and the home by enhancing an individual's knowledge and skills to adapt to the demands of today's rapidly changing society. Special consideration is given to meeting the needs of groups such as low-resource families, youth and families at risk, senior citizens, and young homemakers.

4-H. The mission of the 4-H and youth development program is to create supportive environments in which culturally diverse youth and adults can reach their full potential (Cooperative Extension System 1994). The program strives to develop life skills to help young people become self-directed, productive citizens. The mission also includes empowering adult volunteers who multiply the efforts of professionals.

Community Development. Community development programs focus on improving the physical, economic, social, cultural, and institutional environment in which the people of a community live and work. Leadership development and efforts in revitalizing rural America are major base program areas. Public policy education helps citizens make informed choices about whether or not they want change. Leadership programs are "designed to make people more effective at whatever they choose to do" (Boone 1988).

Interdisciplinary programs involve more than one program area such as leadership, waste management, water quality, food safety, and economic development. Each of the initiatives will have specific goals that seek to improve the quality of life of the clientele served. Although mission statements are written periodically, Extension's educational focus—the emphasis on practical knowledge, a hands-on approach, and a nonformal learning environment—remains the same.

Conceptual Models

How Extension accomplishes its mission is best illustrated by three conceptual models: technology transfer, problem solving, and imparting knowledge, see figure 1-5.

In the **technology-transfer** model, "the answer is provided to the client" (Boyle 1981, p. 12). Knowledge from the land-grant university is disseminated from the campus out to the people in the state. The role of Extension is to link the client to the research. Ideally, the transfer is in response to specific clientele needs, but the impetus for what information is transferred comes from researchers seeking new knowledge. "There is not an attempt to help individuals gain an understanding of the concepts or the background that produced the information, only an understanding of the specific facts and how to use them" (Boyle 1981, p. 12). An example of

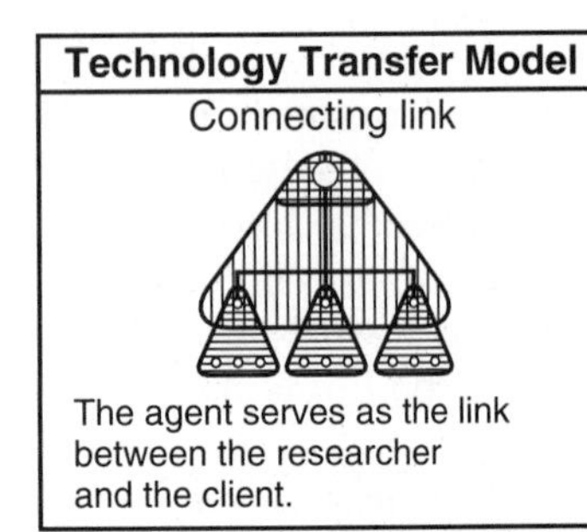

Technology Transfer Model
Connecting link
The agent serves as the link between the researcher and the client.

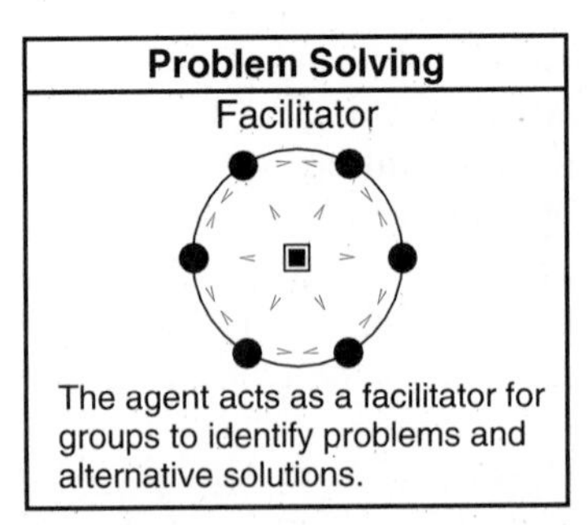

Problem Solving
Facilitator
The agent acts as a facilitator for groups to identify problems and alternative solutions.

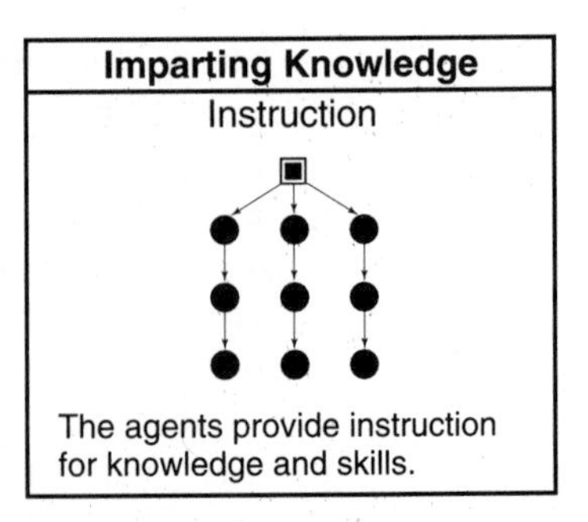

Imparting Knowledge
Instruction
The agents provide instruction for knowledge and skills.

Figure 1-5 Conceptual Models. Gamon, J. (1995). Department of Agricultural Education and Studies, Iowa State University, Ames.

programming using the technology-transfer model would be providing the results of field trials performed by a researcher to a person who wants to know which cultivar will best withstand a cold winter.

The **problem-solving** approach facilitates solutions to group problems identified by clients. Alternative solutions are proposed and evaluated and appropriate action is taken. Here the role of Extension is to facilitate the process of identifying alternatives to problems. The final step in problem solving is reflection on the value of the solution. According to Boyle, "Development programs often begin in very ambiguous situations. Although there may be recognition of a need, the problem is often not well defined nor are priorities established" (1981, p. 9).

In the **imparting-knowledge** model, Extension's mission is to upgrade the skills of individuals unable to attend formal on-campus classes. The focus of this model is on teaching the content of a discipline. A financial management workshop instructed by an Extension specialist is an example of this model in action.

SUMMARY

The Cooperative Extension System is a national educational network that links research, science, and technology to the needs of people where they live and work. The Smith-Lever Act of 1914 established the system as a unique national partnership of federal, state, and local governments. The United States Department of Agriculture is the federal partner. Land-grant colleges and universities are the state partners, and a network of field staff at the local level connects university research to the people or translates "research into reality." Field staff serve as linkages to transfer technology from the researcher to the local level. They also serve as facilitators to empower people to solve their own problems.

The philosophy of Extension is pragmatic and experiential. Extension professionals believe in the ability of people to make their own decisions and the value of learning by doing.

The mission of Cooperative Extension is to enable people to improve their lives and their communities. Cooperative Extension accomplishes its mission by offering practical education for Americans to use in dealing with the critical needs that impact their daily lives and the nation's future.

DISCUSSION QUESTIONS

1. Describe the Cooperative Extension System.
2. Which characteristics of Extension make it unique? Which characteristics are common to other organizations?
3. How might Extension's philosophy be different if its beginning were today instead of in 1914?
4. How might an Extension professional use one of the mission statements with clients?

5. Give an example of a program that would include all three conceptual models: technology transfer, problem solving, and imparting knowledge.

REFERENCES

Blackburn, D. J., ed. 1984. *Extension handbook.* Guelph, ONT., Canada: University of Guelph.

Bliss, R. K., ed. 1952. *The spirit and philosophy of Extension work.* Washington, D.C.: United States Department of Agriculture and Epsilon Sigma Phi.

Bloome, P. D. Spring 1983. Seeking a mature relationship with agriculture. *Journal of Extension* 30(1): 4–8.

Boone, E. 1985. *Developing programs in adult education.* Englewood Cliffs, NJ: Prentice-Hall.

_______ 1988. *Working with our publics.* Raleigh, NC: North Carolina State University.

_______ 1989. Philosophical foundations of Extension. In *Foundations and changing practices in Extension*, D. J. Blackburn, ed. Guelph, ONT., Canada: University of Guelph.

Boyle, P. 1981. *Planning better programs.* New York, NY: McGraw-Hill.

Cooperative Extension System. 1994. *Focus on the future: a strategic plan for 4-H and youth development.* Washington, D.C.: Cooperative Extension System, USDA.

Cooperative States Research Education and Extension Service. May 1995. *Research and education for the 21st century.* Washington, D.C.: Coöperative States Research Education and Extension Service, USDA.

Extension Committee on Organization and Policy. August 1990. *Conceptual framework for Cooperative Extension programming.* Washington, D.C.: Strategic Planning Council, Extension Committee on Organization and Policy and Extension Service, USDA.

_______ 1995. *Framing the future: strategic framework for a system of partnerships.* Champaign, IL: University of Illinois Cooperative Extension Service.

Fiske, E. P. 1989. From rolling stones to cornerstones: anchoring land-grant education in the counties through the Smith-Lever Act of 1914. *Rural Sociologist* Fall:7–14.

Prawl, W., Medlin, R., and Gross, J. 1984. *Adult and continuing education through Cooperative Extension Service.* Columbia, MO: University of Missouri Press.

Rasmussen, W. D. 1989. *Taking the university to the people—seventy-five years of Cooperative Extension.* Ames, IA: Iowa State University Press.

Safrit, R. D. 1990. *Values stratification in the strategic planning process of an adult education organization.* Unpublished doctoral dissertation. Raleigh, NC: North Carolina State University.

Safrit, R. D., Conklin, N. L., and Jones, J. M. 1992. Organizational values of Ohio Cooperative Extension Service employees. In *Proceedings: Symposium for research in agricultural and Extension education.* Columbus, OH.

Sanders, H. C. ed. 1966. *The Cooperative Extension Service.* Englewood Cliffs, NJ: Prentice-Hall.

Sanderson, D. R., ed. 1988. Understanding Cooperative Extension: our origin, our opportunities. *Working with Our Publics*, Module 1. Raleigh, NC: North Carolina State University.

Smith-Lever Act as Amended in 1962. Public Law 83. 83rd Congress. First Session. S 1679. Chapter 157. 1962.

Smith-Lever Act as Amended in 1985. Public Law 99-198. 99th Congress. S 1435. 1985.

Smith-Lever Act of 1914. 63rd Congress. Session II. S 372. Chapter 79. 1914.

Ward, B. 1962. *The rich nations and the poor nations.* New York, NY: W.W. Norton.

White, B., and Brockett, R. G. Summer 1987. Putting philosophy into practice. *Journal of Extension* 25(2):11–14.

CHAPTER

"No race can prosper until it learns that there is as much dignity in tilling a field as in writing a poem."
—Booker T. Washington

The Origin of Extension Work

Key Concepts

- Historical events and needs of American education and agriculture
- Cooperative demonstration work movement
- Major laws leading to the creation of Cooperative Extension
- Extension's role in the land-grant system
- Cooperative Extension Service in the twentieth century

Agriculture: The Nation's Central Pursuit

Throughout its history, Cooperative Extension has encouraged invention, experimentation, and education. The understanding of the origin and success of this organization is based on a knowledge of its history. The documentary, *Agriculture of the United States* (Rasmussen 1975), provides the background for understanding the series of events that led to the creation of this unit of the land-grant university.

In 1783, with the Treaty of Paris, the United States became a free, independent nation. This new nation faced many problems, and one of the major problems related to agriculture. It was an agricultural country—by about 1800, the nation's farm population numbered approximately 4.3 million or 85 percent of all employed persons. Land disposition, restrictions on agricultural trade, the issue of slave labor, and the need for higher yields and better products were only a few of the farm-related problems.

Informal experiments to identify suitable crops and production methods were extensive during the colonial period. Plant-testing gardens and botanical gardens served as sources of seed. The U.S. Treasury encouraged foreign countries to send plants and seeds to establish plant introduction gardens. Information about experimental findings was circulated with equal informality, mostly by word of mouth. Friendly Native Americans taught the colonists how to raise corn, tobacco, and other crops. Squanto is called the first demonstration agent, as he advised the New England colonists to fertilize corn with fish.

Influential planters experimented with different crops. George Washington developed a strain of wheat by careful seed selection. He was a proponent of improved farm equipment and one of the first Americans to raise mules. As President, he urged the Congress to create a national agricultural agency that would diffuse information to farmers.

Thomas Jefferson was known as a scientific farmer who experimented with new crops and developed a number of new pieces of farm machinery. He was a regular correspondent about his belief in agriculture as a way of life and advocated a network of agricultural societies.

A federal Patents Office was created in 1790 with the majority of the patents approved for agricultural use. The first Commissioner of Patents, Henry L. Ellsworth, encouraged a seed distribution and export program. His early work helped expand the idea that led to the creation of the Department of Agriculture. Although there were many efforts to improve agriculture, they were very fragmented at this period of time.

Agricultural Societies

The first systematic attempts to improve agriculture came shortly after the American Revolution, when the agricultural societies began. The movement grew steadily for approximately seventy-five years and reached its peak with about 900 societies around 1861.

Scientific groups such as The American Philosophical Society and the American Academy of Arts and Sciences encouraged agricultural experimentation and led to the creation of societies devoted only to agriculture. The Philadelphia Society for Promoting Agriculture, organized in 1785 by Benjamin Franklin, is believed to be the first such society. The South Carolina Society for Promoting and Improving Agriculture and Other Rural Concerns was organized in the same year. These early societies were composed of professional men who could afford experimentation by studying improvement in other countries and adapting these to American farm conditions. Some societies, such as the Berkshire Agriculture Society organized in 1811 in Massachusetts, promoted the improvement of cattle through livestock shows or fairs by awarding premiums. This idea spread rapidly throughout all the agricultural societies. States began to provide grants for premiums at the fairs to promote better crops and fatter animals, see figure 2–1. Societies such as the Ohio Company for Improving English cattle, established in 1833, promoted the improvement of livestock

Figure 2–1 ***Cattle shows and fairs were used to promote better crops and fatter animals at the turn of the century.***

breeds. The Morgan horse was introduced in Vermont and longhorn and shorthorn breeds of cattle were imported from England.

Several agricultural inventions helped promote the cause of agriculture production. These included the cotton gin, the iron plow, and the mechanical reaper. But these new inventions could not resolve the problem of soil depletion from tobacco, cotton, wheat, and corn production. Many early agricultural educators experimented with manure, lime, and gypsum as a means of improving the soil. Edward Ruffin, the first American soil scientist, published essays on calcareous manures as early as 1821 as agricultural publications became common. Two examples were the *American Agriculturist* and the *American Farmer*. The latter gained prominence through its editor, John Stuart Skinner, who became an advocate for superior crops and livestock.

Agriculture societies grew in popularity and were found at state and county levels. These societies published journals, sponsored programs of education, and occasionally invited a visiting lecturer from a college. Most faltered after the depression of 1825 but the idea revived around 1840 when states again began to make subsidies for premiums to reward agricultural innovations. The movement peaked with the establishment of the United States Agricultural Society with headquarters in Washington, D.C.

The growth of the agricultural societies played an important role in the concern for the education of the farmer. Agricultural societies were successful at sharing new information about farming and promoting the use of better crops or livestock, but the message was slow to be adopted by the multitude of farmers who felt little need for change.

Land Grants: An Idea that Changed America

Many leaders of the 1800s favored the sale of land for colonization. New territories were given to soldiers, while other western land areas were claimed by existing states or sold to land companies, such as the Ohio

Company. The Land Act of 1800 allowed land to be sold directly to settlers, mostly in tracts of 320 acres and included credit provisions. This direct sale of land helped bring about a rapid expansion of the western areas and bring states into the Union. Public lands were also granted to the railroad companies through the Public Railway Act, an act which helped connect the East and the West through a railroad system.

The settlement of frontier land culminated with the signing of the Homestead Act in 1862. This act gave 160 acres of public land to anyone who was the head of a family or over 21 years of age. The title was issued after the settler had made improvements to the land and had lived on it for six months, see figure 2-2. However, new land presented new problems in agriculture production and an even greater need for education of the farmer.

Figure 2-2 ***This homestead is typical of many settled public lands as a result of the Homestead Act. Courtesy of Arkansas Agricultural Experiment Station.***

A COLLEGE FOR THE COMMON PEOPLE

The first American colleges were based on the English University model. The course of study was founded on the classics which prepared students for the ministry, medicine, and the law. But in the nineteenth century, agriculture was the nation's central pursuit, so strengthening agriculture meant strengthening the nation (Boone 1989). A different type of college was needed, one devoted to educating the common people whose lives would not be spent in the professions but in business and trade. Jonathan Turner of Illinois was one of the most vocal advocates for a system of education for the working man. His essay, "A Plan for a State University for the Industrial Classes" published in 1850, contained most of the ideas that later became the Morrill Act of 1862. Thomas Clemson also promoted the study of science and education in agriculture and was instrumental in establishing colleges along the eastern coast. The movement in agriculture paralleled and often crossed the development of science. Speeches, letters, editorials and public dissatisfaction with the plight of the farmer were common.

The first school devoted exclusively to agriculture was established in 1823 as the Gardiner Lyceum in Maine. Others, such as Kings College (later Columbia University) and Harvard College, offered instruction in practical agriculture. Limited attempts were also made to establish experimental farms and analysis laboratories modeled after the great Rothamstead station in England.

Pennsylvania established an agricultural high school in 1855, which, in 1862, became the state's land-grant college, and later Pennsylvania State University. Evan Pugh, first principal of the Farmers' High School and first president of the Agricultural College of Pennsylvania, led the cause in Pennsylvania to establish a "central college with branch schools in each county, investigation stations in each township, and observers on each farm. The college was to train not only farmers and farmer's wives but research specialists and teachers for the rural schools" (Eddy 1957). Pugh was opposed to the manual labor requirement advocated at most of the agricultural schools. He wanted an institution that would be professionally scientific, where agriculture and the mechanic arts would be combined into one college.

Michigan became the first state to establish a college of agriculture in 1855, after many years of pressure from the Agricultural Society of that state. Manual labor was required by law of all agricultural students and continued to be a requirement for forty years. In 1856, Maryland also established a college of agriculture creating a triad of pioneers in the national movement for agricultural and mechanic arts institutions—Pennsylvania, Michigan, and Maryland.

The Morrill Act of 1862

Although several states had passed legislation to open schools to study agriculture, all suffered from the lack of quality teachers, curriculum, and financial resources. There was a general feeling that these colleges would not receive adequate support without the contribution of the federal government (Sanders 1966). Convinced of the public sentiment for such schools, Representative Justin Morrill of Vermont introduced the land-grant bill to Congress in 1857. This bill donated federal land to each state and territory for an endowment to establish:

> *at least one college in each state where the leading object shall be, without excluding other scientific or classical studies, to teach such branches of learning as are related to agriculture and the mechanic arts, as the legislatures of the states may respectively prescribe, in order to promote the liberal and practical education of the industrial classes in the several pursuits and professions of life.*
>
> *(Eddy 1957, p. 31)*

President Buchanan vetoed the bill, citing current economic conditions, the danger of land speculators, the inclusion of science and liberal studies, and the fact that Congress did not have the constitutional right to appropriate money for education. Morrill introduced a similar bill three years later that added provisions for teaching military tactics and President Abraham Lincoln signed the Morrill Act into law in 1862.

Iowa became the first state to accept the conditions of the Morrill Act, with Vermont and Connecticut following in 1862. Fourteen states adopted the legislation in 1863. After eight years, 16 states had agreed to establish state-sponsored teaching of agriculture, mechanic arts, and military tactics.

It was also during 1862 that the Organic Act was passed by Congress creating the U.S. Department of Agriculture. The new organization operated as a division of the Patents Office and functioned primarily as an office of communication. Annual reports including research were published and distributed to farmers. In 1889 the Department began publishing Farmers Bulletins, which were later reformatted and became the *Yearbook of Agriculture.*

The Struggle. The Morrill Act was an educational revolution but it took nearly a hundred years for the idea to fully develop. The first twenty years were nothing short of a struggle.

The word *college* in 1862 was synonymous with practically all instruction above the level of the "common" schools. In most communities public high schools did not exist and private academies furnished all but basic instruction. While there was great concern over equal educational opportunities for the "industrial classes," those who were educationally, socially, or economically underprivileged, there was also a more realistic distress over agricultural conditions. Farmlands were being stripped quickly of their fertility and usefulness. The soil was rapidly exhausted of plant food because of inadequate conservation knowledge. The Morrill Act was seen as the answer for farmers and farming, see figure 2–3.

But in the years to come, controversy and bitterness characterized the fight for colleges in almost every state. Many community leaders were in-

Figure 2–3 ***"We do not want science floating in the skies; we want to bring it down and hitch it to our plows."* This was a common sentiment of discontented farmers with early agricultural colleges and experimental work. Arkansas Agricultural Experiment Station. Courtesy of Arkansas Agricultural Experiment Station.***

terested in the financial aspect of the land-grant school, while other communities fought over the location of the new institutions. Two distinct factions developed concerning the purpose of the land-grant schools. The "narrow gauge" argued that science should be applied and taught by hands-on laboratories, shops, or experimental plots. The "broad gauge" supporters argued for a broad liberal curriculum with pure, theoretical science and teaching from books or lectures (Sanderson 1988).

In spite of centuries of farming, there was not an adequate body of knowledge from which the faculty could offer adequate instruction. There were strange combinations in curriculum such as a course in "how to plow" along with one on "mental and moral philosophy." The teaching of mechanic arts centered around shop practice and practical training. Displays of the college shop—castings, forgings, and machinery—were used to convince the public of the value of the training. To find something to teach in agriculture, the colleges relied on the operation of a model farm. Here, the few students that were enrolled would have the opportunity for observation and practice. The model farm became one of the most important parts of the college.

The idea of manual labor was adopted as a requirement for almost all of the new colleges in agriculture. The agriculture students spent time clearing fields, caring for livestock, and helping to plant and harvest crops on the model farm. The students in mechanic arts were expected to

*Quoted in Strausberg, F. (1989). *A Century of Research:* Centennial History of Arkansas Agricultural Experiment Station. University of Arkansas Agricultural Experiment Station: 9.

spend an equal number of hours in the shop as well as help with outdoor efforts to build and beautify the campus.

The institutions did not grow for many years. One of the chief enrollment problems was the inability to interest students in agriculture courses. Agriculture meant simple farming to many. Most young men did not see the reason for spending four years in college studying a subject learned by practical farm experience in much less time. There was considerable suspicion and resistance to "book-learning" but "book-farming" was just not accepted.

Criticism from the majority of farmers mounted during this time also. The dissatisfied farmers felt they had been mislead. They believed, because of the over-emphasis on agriculture and discussion of the Morrill Act, that the colleges would be devoted solely to farming. They were amazed to discover that the colleges offered other courses that the farm students were expected to study. The new education, according to them, should be farm centered, not science centered.

Because of the public interest in education, editors of newspapers and journals and churches added their criticism about the colleges. One journal author called them schools where "hay seeds and greasy mechanics were taught to hoe potatoes, pitch manure, and be dry nurses to steam engines" (Eddy 1957, p. 73). The criticisms were so severe and widespread that the struggling new educators were forced to spend all their time defending their new and undefined practices.

Justin Morrill continued his interest and frequent efforts to gain more funds for the colleges, but it was not until 1890 that Congress would act upon this bill again.

The Hatch Act of 1887

The land-grant colleges contributed greatly to the development of the laboratory method; the idea of experimentation had started early in these colleges. Maryland had various agricultural experiments under way as early as 1858. Scientific investigation was spurred by the constant demand for practical application. The universities in Europe had a profound influence on this experimental work. Wilbur O. Atwater was so convinced of the need that he helped establish the first American experiment station at Wesleyan University Middletown, Connecticut, in 1875. Atwater became the first director of this station and later the first director of the federal office of experimental stations. California and North Carolina were the next states to establish organized experimentation work in a station-type program in 1877. Maryland was the first state to have some kind of organized experimentation work on the college farm. The agricultural experiment stations marked the start of the direct application of science to the problems of agriculture, see figure 2–4.

The experimental farms soon became the focus of the principle concerns of farmers. The farmers wanted to know what was best and most profitable, how to control disease, how to feed and fatten livestock, and

Figure 2–4 ***Recommendations from the first experimental farms dealt with control of diseases and insects. From* The People and the Profession.**

how to grow crops to obtain the best returns. The stations were hard pressed to find immediate answers and too understaffed to deal with more fundamental problems.

This concern provided a substantial reason for representatives of the land-grant colleges to begin discussing the need for federal aid to the agricultural experiment stations. Because of the national significance, legislation was prepared and introduced into Congress. Professor Seaman Knapp from Iowa led a delegation in addressing the Commissioner of Agriculture, arguing that farmer's work needed the assistance of the government. Knapp explained that a station was needed in each state because of the diversity of climate, multiplicity of problems, and the attention needed for the acclimation of foreign seeds. He proposed uniting these stations with the agricultural colleges to avoid duplication of buildings, apparatus, and personnel. Support for such work was advocated by agricultural societies, the Grange, and other groups interested in agricultural education or research. A number of bills in support of the experiment stations were introduced from 1862 to 1887. After much debate, a bill sponsored by Representative William H. Hatch of Missouri, chairman of the House Committee on Agriculture, and Senator J.Z. George of Mississippi in support of such experimental stations was passed into law.

The Hatch Act of 1887 provided for a department to be designated and known as the agricultural experiment station in each of the colleges established under the Morrill Act. Its purpose was

> *. . . to aid in inquiring and diffusing among the people in the United States useful and practical information on subjects connected with agriculture and to promote scientific investigation in experiments respecting the principles and applications of agricultural science.*
>
> *(Prawl 1984, p. 18)*

The stations were required to publish periodic bulletins or reports of projects and make them available to the public. Funds were provided to pay for the expense of conducting these experiments and the printing and distributing of the results. With the Hatch Act the federal government entered into a systematic and cooperative relationship with the colleges, though the institutions were to remain distinctly state organized and sponsored. Local initiative was maintained but national interests were recognized. The passage of the Hatch Act immediately stimulated the growth of the experiment stations and the number of staff members employed for experiment station work doubled within ten years.

The Morrill Act of 1890

Senator Morrill had been struggling for almost twenty years to get more help for the new colleges. He introduced twelve bills between 1872 and 1890 to further the endowment of the land-grant colleges but all were defeated. Finally in 1890, Morrill introduced an amendment for an additional endowment for instruction at the land-grant schools that succeeded. In this legislation, states that were maintaining separate colleges for the different races had to propose a just and equitable division of the funds to be received under the act. The states that had used the 1862 funds entirely for the education of white students were forced either to open their facilities to black students or to provide separate facilities for them, see figure 2–5.

Federal support of the land-grant colleges became law with the passage of the Morrill Act of 1890. The act spurred the establishment of sixteen black land-grant colleges throughout the South and gave land-grant status to Tuskegee Institute. The 1890 bill also made provisions for teaching some of the arts and all of the sciences. It is believed that this provision contributed to the development of universities rather than colleges limited to agriculture and mechanic arts.

The second Morrill Act serves as an important milestone in the history of the land-grant colleges. With the addition of federal funds, the institutions that had been struggling to survive since 1862 were now ready to become progressive segments of American higher education. While the

Figure 2-5 ***The establishment of 1890 land-grant colleges provided for equitable demonstration work with black farmers. From*** **The People and the Profession.**

Morrill Act intended for "equitable" distribution between white and black colleges, this did not happen. Most of the institutions founded in 1890 were forced to offer college preparatory courses for black students. However, this helped to develop teacher training programs at these institutions, which may have been the greatest contribution of these colleges as this helped to address basic educational needs of blacks at this time in history.

THE IDEA OF EXTENSION EDUCATION

The Lyceum movement in education was very popular prior to the American Civil War. This method was similar to a community meeting where people gathered to hear lecturers. Additional self-improvement came to those who attended the assemblies on Chautauqua Lake (New York) for a program of instruction, recreation, and entertainment. This was replaced by the Chautauqua system of correspondent education, a course in home reading. Correspondence courses were added in 1883.

The idea of "university extension" or off-campus, noncredit courses, began in England and was introduced in the United States through city libraries. By 1890, extension courses were frequently offered. The agricultural college at Rutgers (New Jersey) was one of the first to offer an agricultural extension program in 1891 with off-campus courses on soils and crops, feeding plants, and animal nutrition. The Ohio University was one of the first to offer on-campus instruction for farmers in a noncredit, nonexamination course.

Farmer's Institutes and "Movable Schools"

During the 1890s and into the first decade of the twentieth century, the colleges' major device to "extend" the boundaries of the campus was through the Farmer's Institute (Fiske 1989). This extension was provided by experimental station staff as part of their job to make the information of the laboratory and experimental field available to anyone who desired it or could use it. The first Farmer's Institute was held in Springfield, Massachusetts, in 1863 by the State Board of Agriculture. By 1899 all but three states were conducting Farmer's Institutes on a regular basis. It was estimated that some 2,000 institutes were held that year with more than one-half million farmers, women, and youth in attendance, see figure 2–6.

Topics of interest to the farm wife and youth were soon added to the Farmer's Institutes. In 1890, women speakers were taking part in institute programs such as cooking schools in Minnesota. In 1895, a separate section for women was organized in Michigan. By 1898 Illinois women had formed a domestic science association and worked with the men in planning and carrying out the institutes.

Figure 2–6 ***Farmer's Institute provided farmers the opportunity to see the latest development in agriculture. Courtesy of Special Collections Division, University of Arkansas Libraries, Fayetteville.***

Early extension efforts for rural black families began under the leadership of Booker T. Washington, the founder of Tuskegee Institute. Washington led tireless efforts to help rural blacks improve their farms and homes using many methods to deliver useful and practical information, even the pulpits of churches. He organized a Farmer's Institute at Tuskegee in 1890, which was attended by over 500 black farmers. In 1896 he convinced George Washington Carver to come to Tuskegee and organize a Department of Agriculture. Carver had the same zeal for helping rural farmers and continued to try different methods of teaching. Carver's most popular idea was that of the "movable school," introduced in 1899. He developed a plan for a mule-drawn wagon that carried farm machinery, seeds, dairy equipment, and other materials to demonstrate improved methods of agriculture to black farmers. This wagon was known as the "Jessup Wagon" in honor of the New York philanthropist who financed the purchase of the first wagon.

In 1903 Perry Holden of Iowa got the idea to use railroad trains for educational work with farmers. College lecturers would travel by train equipped with corn specimens, charts, bulletins, and demonstration materials. At each stop, farmers could hear lectures, secure publications, and walk through the train to study the exhibits. This idea was so successful that many other colleges and railroads began to use this method, see figure 2-7.

The popularity of university extension worked hardships on the already hard-pressed college faculties. Many of those working in experiment stations were carrying class loads as well. Now they were asked to travel throughout the state to conduct demonstrations and write bulletins, reports, and articles for newspapers and journals as well as attend frequent farmers' meetings.

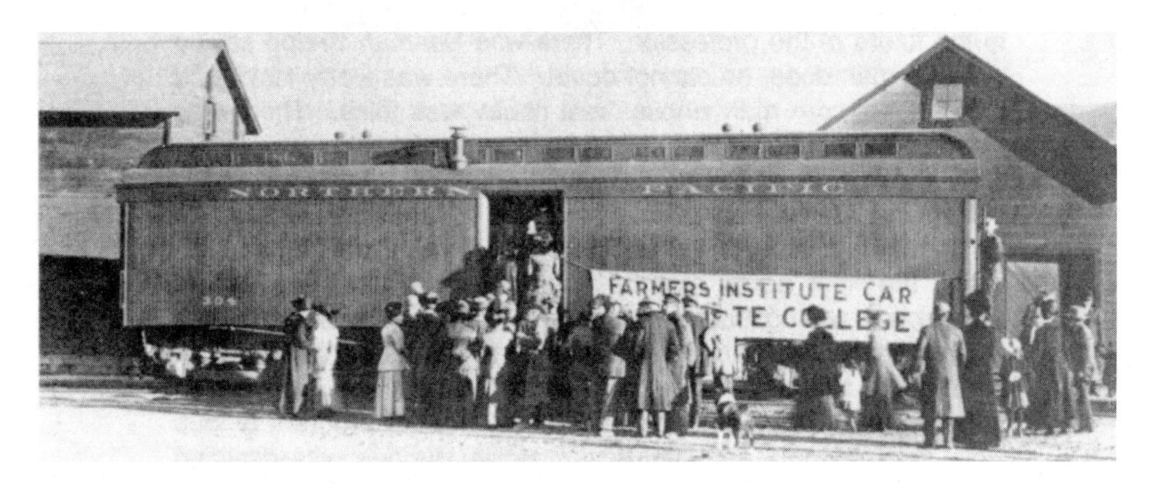

Figure 2-7 ***Traveling railroad exhibits brought the latest innovations to the farmer in the 1900s.*** *From* **The People and the Profession.**

Seaman A. Knapp: A Social Pioneer for Extension Education

If all great organizations and institutes are but the lengthened shadow of a man and an idea, then Extension education is the shadow of Seaman Knapp and his idea of teaching by demonstration. Seaman Asahel Knapp was born December 16, 1833, at Schroon Lake, New York, see figure 2–8.

Knapp possessed great physical and spiritual endurance, a rugged constitution, and a strong sense of community service. He had a distinguished career as a minister and school educator, but was forced to leave the teaching profession after a serious knee injury. During this period of ill health from 1869 to 1875, he regularly read reports of scientific research and experimentation about the application of science to agriculture. It was the interest in these experiments that helped him return to farming with renewed interest after his recovery.

Knapp decided to enter the pig business and try this new science. He experimented with new practices, lineage, and balanced diets and shared his experiences in *The Farmer's Journal*. He became a dealer in thoroughbred swine and supplied brood sows and boars to the farmers who sold their animals for slaughter. He took advantage of every opportunity to promote better production. He organized the Plough Handle Grange and talked to county fair audiences and agricultural conventions and societies. He organized the Iowa State Improved Stock Breeders Association, became president of the Benton County Agricultural Society, and served as director of the Iowa State Agricultural Society. He also assisted an organization that published the *American Poland-China Record Book* to validate

Figure 2–8 ***Seaman A. Knapp, the originator of farm demonstration work.*** ***From*** **Taking the University to the People: Seventy-five Years of Cooperative Extension.**

breeding lines. He was tireless in advocating scientific farming and motivated farmers to change their ways by using some of the same persuasive zeal he had learned in the pulpit.

Knapp became the editor of *The Farmer's Journal*, also called the *Western Stock Journal*. He felt compelled to provide farmers with more practical facts to prevent trial-and-error farming. Knapp teamed with H. W. Wheeler to organize a series of livestock sales and auctions, bringing large numbers of stock together so that the farmers would have a good selection. He also provided credit through a bank he opened.

Knapp was a social pioneer noted for his persuasive personality, sincerity, honesty, and desire to benefit and instruct. He became an outspoken supporter of experimental farms. He attended conventions, Farmer's Institutes, and other meetings with leaders in agriculture and was influential in establishing institutions to provide the education, periodicals, stock, equipment, and credit that helped the rural population rise above subsistence farming.

Agriculture Demonstrations

In 1879, Knapp was elected professor of Agriculture at Iowa State College where he gained a reputation for the practical methods of teaching. He later became president of the college. He instilled pride and honor in the study of agriculture and hard work even while many students were ridiculed as "hay seeds." Knapp's interest was in promoting farm experimentation. He published a series of publications in 1883 detailing his experiments on dairy cows, feeder pigs, and crop varieties.

Although he annually requested liberal appropriations for experimental farming, Knapp had little luck in receiving state appropriations. But the idea of a national appropriation for each state for agriculture experimentation was being promoted by the Society for the Advancement of Agricultural Science and a new group of The Teachers of Agriculture (organized in 1880), of which Knapp was a member. Knapp drafted the bill for federal aid to agricultural experiment stations, which in 1887 became the Hatch Act.

In 1896 Knapp resigned as president of Iowa State College to manage a company planning to colonize over a million acres in Louisiana. He was 69 years old. After many trials and errors, Knapp resorted to demonstrations to convince settlers that the land was fertile enough for farming. Knapp believed that rice would be the best crop for the area, but the varieties grown locally were not commercially satisfactory. He traveled to Japan, China, and India to select better varieties of rice to introduce in the United States.

Knapp was later appointed to a position in the Bureau of Plant Industry in the USDA. The Bureau was especially interested in encouraging farmers to adopt better farming practices. Knapp suggested the demonstration farm model and was placed in charge of the program. These farms were designed to show how to increase yields of the standard crops. It was through this work that Knapp learned the power of agriculture demonstration. His philosophy is best described in one of his most often-quoted sayings,

> *What a man hears, he may doubt. What he sees, he may possibly doubt. But what a man does himself, he cannot doubt.*
>
> *(Bliss 1952, p. 240)*

But Knapp feared that the farmers would think any individual could produce good results if the government was backing him. So the idea of a demonstration farm without government subsidy was conceived.

E. H. R. Green, president of Texas-Midland Railway, was interested in the "community demonstration farm" and invited Knapp to Terrell, Texas. Knapp, who was determined to test his idea of locally owned demonstration farms, convinced the community of Terrell to raise $900 to insure against crop loss. Walter C. Porter agreed to farm about seventy acres according to Knapp's instructions and to keep records of costs and yields. The Porter Community Demonstration Farm attracted much attention, especially when Porter reported that he had made $700 more as a result of farming by Knapp's recommendations.

In 1903 Knapp received a special allotment of money to establish cooperative demonstrations for cotton farmers facing ruin from the boll weevil. About seven thousand demonstrators were enrolled by 1904 and twenty-five to thirty field men called "special agents" were appointed that year to work with these demonstrators, see figure 2-9. Cotton yields on

Figure 2-9 ***The extension agent's working conditions were often primitive and arduous. From*** **The People and the Profession.**

demonstration farms were twice as large as the average yields on farms in the same localities where the demonstration methods were not followed. The advantage of following the recommendations of the field man set an example soon followed in many other areas of Texas and Louisiana. The idea was so popular that farmers and businessmen petitioned Knapp to have a man work exclusively in Smith County, Texas. W. C. Stallings was the first county agent appointed by the USDA in November 12, 1906.

While Knapp was working in Texas, the extension movement under way at Tuskegee Institute in Alabama under the leadership of Booker T. Washington and George Washington Carter was also gaining popularity. Knapp proposed that Tuskegee and the General Education Board unite forces for cooperative demonstration work. The General Education Board was created by John D. Rockefeller from Standard Oil Company funds. One of its purposes was to raise the level of education in the South. The General Education Board agreed to help finance the new Farmers Cooperative Demonstration work. The proposal that Tuskegee and the General Education Board unite was an effort to jointly finance the programs. Thomas Campbell was appointed as the first black USDA field agent on November 10, 1906, and was placed in charge of the demonstration wagon, see figure 2–10. J.B. Pierce of Hampton Institute of Virginia was appointed a few days later. By 1912, there were thirty-two black demonstration agents working with thirty-five hundred black farmers and an additional ten to fifteen thousand black farmers under Knapp's direction. This working relationship contributed to great improvement in the

***Figure 2–10** Thomas Campbell was the first black field agent on appointment with USDA. Tuskegee University Archives.*

economic and social conditions of agriculture and race relations in the South (Bailey 1945).

The cooperative demonstration work was so successful that it led to the hiring of agents in many southern states with a field organization consisting of a state agent, district agent, and county agent. J. A. Evans was state agent for Louisiana and Arkansas. W. C. Proctor was state agent for East Texas and J. L. Quicksaul for West Texas. W. D. Bentley was state agent for Oklahoma and R. S. Wilson for Mississippi and Alabama. All of these men lead distinguished careers in Extension service, which are noted in *Recollections of Extension History* by J. A. Evans. The idea of a local field agent holding a joint appointment with a college and the USDA for demonstration work, modeled after Clemson College, set the pattern for the legislation that would follow.

Boys' and Girls' Clubs

Boys' and girls' club work, now known as 4-H clubs, developed from the idea of education in the public schools and a desire to teach the latest agricultural practices through youth. Many educators, including Liberty Hyde Bailey of Cornell University, Ithaca, New York and Albert B. Graham, Superintendent of Schools for Springfield Township in Ohio, questioned the relevance of the public schools' instruction for rural youth.

Bailey used funds appropriated in New York for Extension work to distribute nature-study leaflets to the rural schools. Graham popularized the idea of study outside the classroom. He also tried to model the idea of vocational education from the urban schools that included mechanic arts and typing. The need for rural youth to learn about agriculture and household management was his concern. In 1901, Graham established experimental corn clubs outside of school hours by holding meetings on Saturday mornings while the parents were shopping in town. He received assistance from the Ohio Agricultural Experiment Station because the Agriculture Dean, Thomas S. Hunt, thought that Graham's agriculture clubs would be a great opportunity to distribute information to the farm community. Graham's work was so successful that he became the first Superintendent of Extension at Ohio State University in 1905.

Similar work was taking place in Illinois, led by O. J. Kern and Will B. Otwell. Otwell was the president of the Farmer's Institute in Macoupin County, Illinois. Since farmers were not interested in attending the meetings, he concentrated his efforts on the farm youth. Otwell offered a one-dollar premium for the best corn yield and 500 boys requested seed corn for the contest. In 1901, Otwell's corn-growing contest attracted 1,500 boys. The idea spread as manufacturers offered prizes like cultivators to the winners at the St. Louis Exhibition in 1904, which attracted 50,000 entries. This became one of the most nationally advertised efforts of boys' club work and within a few years corn-growing contests and other agricultural projects were found in many states, see figure 2–11. The idea of

***Figure 2–11** Corn-growing contests and other agricultural projects were popular ways to spread innovations among youth. Photo Courtesy of CSREES/ USDA.*

awarding premiums for agricultural projects at county and state fairs became very popular. Girls' clubs had similar competitions in flower growing and home gardening.

In the South, boys' and girls' clubs were supported by the General Education Board, an organization established by John D. Rockefeller in 1902 to promote education in the United States. It is believed that the first work using USDA funds for youth work originated as a boys' corn club organized in 1907 in Holmes County, Mississippi, by W. H. Smith, the school superintendent. Knapp was interested in this approach and appointed Smith a "collaborator" in the USDA and gave him free mailing to conduct his work.

Girls' canning clubs were believed to have been started in Aiken County, South Carolina, in 1910 by a rural school teacher named Marie Cromer. Upon attending a school improvement meeting with representatives of USDA, she organized the girls' club and encouraged members to grow 1/10 of an acre of tomatoes. Representatives from USDA helped with this project. Similar work was organized by Ella G. Agnew in Virginia under the sponsorship of the General Education Board. A number of women trained in home economics assisted with this work and became known as home demonstration agents.

Until federally sponsored corn-growing contests appeared in 1907 in Mississippi, practically all club activity took place in the schools or under teacher supervision. Farmer's Institutes conducted special sessions for youth and encouraged schools and other organizations to initiate clubs for boys and girls. This was found to be the best method for introducing new techniques and crop varieties to the farmers. The interest in boys' and girls' club work helped local education leaders to improve the rural school. Two Iowa leaders, Jesse Field and O. H. Benson, introduced the subjects of agriculture and domestic science into the rural school as a course of study. Benson is also credited with the idea of the three-leaf clover as the club emblem. A Nebraska association is credited with the development of the hand, head, and heart parts of the 4-H creed. The health "H" was added years later.

The partnership of cooperative work between county officials, state land-grant colleges, and the federal government that started in Mississippi became the model that was generally approved for Extension work with youth.

Demonstration Work with Women

Attending farmer club and agricultural society meetings was popular among farm wives in almost all states. As early as 1903, more than fifteen states were offering "institutes" especially for women. Early organizations for women in the Midwest, called "domestic science associations," were formed to teach better methods in the home and to promote domestic science in the schools. Other names for these organizations included neighborhood study clubs, homemaker clubs, farm women clubs, and home bureaus.

The home demonstration agent was expected to be an excellent cook, a high-class seamstress, and a scientific dietician (Martin 1941). Many agents gained standing and prestige in the community by being able to prepare simple nutritious foods in the most helpful and practical way. Teaching canning to young girls was a back-door approach to teaching cooking to the mothers. Women agents were also carpenters and cabinet makers. Early reports record construction of ice refrigerators, fly traps, ironing boards, and kitchen cabinets as part of the home demonstration program. The agent was also expected to be a gardener, orchardist, and farmer. Home demonstration agents worked for home sanitation, beautification, the elimination of contagious diseases, and encouraging thrift, see figure 2–12.

The first home demonstration clubs associated with Extension in the South developed from the girls' tomato clubs. As mothers became involved with their daughters in the tomato clubs, the home demonstration agents took advantage of the opportunity to demonstrate improved methods of housework. Agents employed in other regions of the United States encouraged the development of county Extension organizations to present educational information at training schools. Marie Cromer of South Carolina was appointed home demonstration agent on June 3, 1910, and Ella G. Agnew on August 16, 1910. Both had started work with girls' tomato clubs. The first black home demonstration agents, Annie Peters Hunter of Oklahoma and Mattie Holmes of Virginia, were appointed in 1912.

Figure 2–12 Home Demonstration agents were part of early Extension work for women. Cooperative Extension Service, University of Arkansas.

The success of this method of teaching rural America gained support for a national system of Extension work. The agricultural colleges and experiment stations requested federal funds for an Extension unit of the land-grant system through the American Association of Agriculture Colleges and Experiment Stations (AAACES) committee on Extension, led by Kenyon Butterfield. President Theodore Roosevelt's County Life Commission, chaired by Liberty Hyde Bailey, recommended a national Extension system to help educate the rural population. After thirty-two bills and several years of debate, Congress passed the Smith-Lever Act of 1914. The most vocal supporters in the great debate were the sponsors of the bill, Representative Asbury Francis Lever of South Carolina and Senator Hoke Smith of Georgia, who wished to extend Knapp's philosophy of farmers' cooperative demonstration work throughout the rest of the nation. This act created a third component of the land-grant colleges with a cooperative funding arrangement and working relationship that is now replicated around the world.

THE SMITH-LEVER ACT

The Smith Lever Act extended the benefits of federal aid to those colleges established under the acts of 1862 and 1890. Its purpose was to:

> *. . . inaugurate, in connection with these colleges, Agriculture Extension work which shall be carried on in cooperation with the United States Department of Agriculture . . . in order to aid in diffusing among the people of the United States useful and practical information on subjects relating to Agriculture and Home Economics, and to encourage the application of the same.*
>
> *(Eddy 1957)*

Section 2 of the act further specifies that this work would consist of the giving of instruction and practical demonstrations in agriculture and home economics and related subjects to persons not attending colleges, and giving information through demonstrations, publications, and otherwise.

When the Smith-Lever Act became effective on July 1, 1914, farmer's cooperative demonstration work was being conducted in fifteen states. There were 1,151 employees of which 279 were home demonstration agents; 53 of these agents were black men and women. In twenty other states, the office of Farm Management had perfected cooperative arrangements with the agricultural colleges and 113 county agents were employed (Evans 1938). A States Relations Service was established to handle the administration of the Smith-Lever appropriations. This bureau united Extension, Experiment Station work, agriculture instruction, home economics, Institutes, and Farmers' Cooperative Demonstration Work. It was not long until many questions of policy, function, and power of the Department of Agriculture and the colleges arose in the conduct of Extension work under the Smith-Lever Act. The Smith-Lever act was designed, in part, to eliminate much of the duplication of Extension efforts among the colleges, the USDA, and other government agencies by creating one organization for this work. The Act's broad charter and interpretation was almost immediately disputed, with the USDA and the agriculture colleges disagreeing about the role of the county agent. With the aid of Secretary of Agriculture David Houston, a cooperative agreement was reached in 1916 in which the itinerant teaching was maintained by the local agent with oversight of the federal funds by USDA. This Memorandum of Understanding outlined the responsibilities and clarified the working relationship of the partners involved in implementing the Smith-Lever Act, see figure 2–13.

With the agreement in place, the old office of Farmers' Cooperative Demonstration Work became the Office of Cooperative Extension Work for the South and the Farm Management Demonstration Work became the Office of Cooperative Extension Work for the North and West. Both offices operated under the States Relations Committee.

Numerous laws have been passed relating to Cooperative Extension work since the Smith-Lever Act. Perhaps the most sweeping legislation occurred in 1953 with the passage of Public Law 83. This law consolidated and codified ten separate laws relating to Cooperative Extension work that had been enacted from 1914 to 1953. This law simplified the appropriations process and introduced a new formula for funding. The language of section 2 broadened the provisions of the original Smith-Lever Act to read "the giving of instructions and practical demonstration in agriculture and home economics and subjects related thereto." The phrase, "and subjects relating thereto," was added to make certain the new legislation authorized all Extension activities, such as 4-H club work, education in rural health, and similar aspects of Extension programs (Sanderson 1988). Numerous other legislative actions have impacted Extension work throughout its history. In the most recent years, Congress has provided funding for educational programs for target populations. Some examples include the Expanded Foods

State Responsibility	Resulted In
• Establish an Extension Service within each state.	• Separate unit of land-grant university.
• Appoint an Extension Director who is responsible for all extension work in the state.	• Director and other personnel responsible for fiscal management.
• Handle funds for state extension work.	• Agents in agriculture, home economics, and 4-H.
• Responsible for the educational work in agricultural, home economics, and 4-H.	
USDA Responsibility	**Resulted In**
• Establish a Federal Extension Service within the USDA.	• Secretary of Agriculture.
• Appoint a Secretary of Agriculture in charge of the federal program.	• Personnel responsible for the Smith-Lever funds.
• Provide leadership for federal educational programs.	• Federal program leaders in agriculture, home economics, and 4-H.
• Coordinate the educational programs of the USDA.	
• Serve as liaison between the state and federal programs.	
Mutual Agreements	
• Joint planning and approval of educational efforts.	
• An annual plan of work.	
• Joint appointments of all state and county extension personnel.	
• A statement outlining joint cooperation would appear on all printed matter.	

Figure 2-13 Memorandum of Understanding.

and Nutrition Educational Program, Pesticide Applicator Training, and Renewable Resources. The scope of Extension work was broadened in 1985 to include "practical applications of research knowledge" and allow for the federal partner to conduct educational programs on its own initiative under the direction of the Secretary of Agriculture.

Extension Changing, but Unchanging

While the rural population was almost 54 percent of the total population in 1910, it is less than 25 percent today. The farm population comprised 35 percent of all Americans when the organization was created in 1914. Today it is about 1.8 percent. The average farm was 139 acres, compared to 468 acres in 1993 (USDA 1993). Today, almost one in three Americans are nonwhite and 13 percent are over age 65. One in four Americans live in a single-parent household with husband-wife households accounting for only 14 percent of all households. About 40 percent of American children live in poverty and new American jobs are most likely to be in the "service" sector (Sanderson 1988, p. 38). But, the wisdom of the investment provided in the Smith-Lever Act has been confirmed repeatedly in the past decades (Vines and Anderson 1976) because the Cooperative Extension Service has constantly changed to meet the shifting needs of the people.

When World War I began, Extension mobilized to spearhead the nation's food production effort. Emergency funds and additional agents were added. Large-scale educational programs were launched encouraging crop and livestock production, food production and preservation, and clothing conservation projects. The farm depression of the 1920s created a new shift in direction. The emphasis changed from production to economic concerns, efficiency in farm operations, and improving the quality of rural life. Emergency funds were eliminated. Unable to hire professionals, Extension agents trained volunteers who contributed significantly to Extension work. More important, the volunteer approach also stimulated rural leadership development, which many regard as one of Extension's major contributions to rural life.

During the 1920s, Extension also was active in helping organize farm cooperatives to facilitate the group purchase of fertilizers, feed, and other supplies, as well as for the sale of crops and livestock. Many of the cooperatives that Extension agents helped organize during that period continue to operate today.

Group discussions on economic affairs emerged as a new approach during the farm depression. This approach proved successful as more farm families became active participants in county, state, and national public affairs.

Dealing with emergency needs such as drought, availability of agricultural credit, and other problems helped create programs such as the Farm Seed and Loan program. Cooperative Extension was called on to operate many federal and state programs, such as soil conservation or rural electrification, that emerged in the post-depression years.

Home economics programs in the early 1930s were geared toward family self-sufficiency. Typical was a nationwide effort to establish community canning kitchens as an intensive food production and conservation effort.

Cooperative Extension became the single federal agency having a direct educational link with rural America. Because of this relationship, Extension again carried an extra burden during World War II. Among its missions were investigating requests for agricultural deferment from the draft, helping with price control and rationing programs, managing the emergency farm labor program, and promoting increased food and fiber production and conservation.

Extension was given another new challenge after the war. The Bankhead-Flannegan Act of 1945 specified intensified county-level development of the Extension system. The Research and Marketing Act also expanded Extension work and led to increased work with consumers in urban areas. Later, through the Farm and Home Development program, Extension focused on farm management, public affairs, and marketing.

Major advances in agriculture that began in the 1940s created the need for new programs to encourage their adoption. Development of hybrid grain varieties, chemical control of agricultural pests, and new soil tillage and fertilization procedures are only a few of the practices advocated. As research and technology raised production capacities in agriculture, Extension broadened its perspective to include other areas such as community development and family living.

The complexity of modern living resulted in new problems and challenges during the 1960s and 1970s. Cooperative Extension responded with many new or expanded programs: programs for low-income and minority groups, programs for migrant workers, and vastly expanded urban programs. Special funds were appropriated for expanded programs such as foods and nutrition, integrated pest management, energy, pesticide applicator, sea grants, rural development, urban gardening, and others that greatly expanded the knowledge base and expertise within the organization and increased Extension's role with nontraditional audiences. These new initiatives also created changes in the staffing patterns of the organization.

The economic plight of farmers created a major shift in programs in the early 1980s. Stress management and farm business management were necessary coping skills that were taught. Efforts to increase programs in 4-H, family living, community resource development, and natural resources were pursued. Extension shifted its focus to broad initiatives and issues, with agents serving as facilitators of local problem-solving groups. Programmatically, issues of the environment along with social and economic changes in communities have created interdisciplinary approaches to problem solving and program delivery in the 1990s.

Extension has changed more in methodology than philosophy throughout its history. The organization built its success and operated for many years with the image of a county agriculture agent making farm visits to render technical advice to the farm producer. The home economics agent, first called the home demonstration agent, demonstrated the latest methods of food preservation or home improvement. The country youth raised calves, pigs, and horses to show in the county fair. The whole family worked to grow prize-winning garden produce. This is the image immortalized in the famous Norman Rockwell painting *The County Agent.*

Today, when many people think of a county agent it is in an urban or suburban scenario—not barnyard, but backyard—and the objects in question are roses, shade trees, and lawns (Ode 1989). The home economist addresses consumer issues or family life, and the county agent is not always a man. Country youth are still involved in Extension programs, but many youth participants live in the inner cities. These youth may work to improve the community, learn consumer skills or computer technology, or study about teen concerns that effect self-esteem. The demonstration method is still relevant, but it is being replaced by new methods of information dissemination in order to reach a greater audience. Radio, television, interactive video, and classroom satellite broadcasts are used along with written fact sheets and group meetings.

While the family unit is still valued, educational efforts today are more likely to reach the members of a family individually. There are very few examples of the "total family" approach today unless it is found in a 4-H club, where the parents serve as leaders and the child is a member of the club.

Organizational stress has become commonplace within Extension as states are forced to downsize staff. These reductions have been a result of the recession in agriculture in 1980s, attempts to reduce the budget deficit, the impact of changing demographics, and unrealistic budget projections in many states. Technological changes have forced Extension to redirect resources and train its employees to teach using these new delivery methodologies. In many states the phrase, "whatever happened to the county agent?" is a realistic one (Ode 1991). The "county agent" no longer exists in many areas as agents have assumed multi-county jurisdictions or positions have been cut. Community programs have been replaced by regional programs taught by a specialist, rather than a generalist.

The organization has met many challenges in the past and will continue to adjust to meet the future. Regardless of adversity, the organization has remained focused on its mission of "taking the university to the people." Retired agricultural historian, Wayne D. Rasmussen (1989) states,

> *no other country has focused such attention on the practical (applied) dimension of education by extending and applying the knowledge base of our land-grant universities to the laboratories of real life where people live and work, develop and lead. Extension has been copied by many countries, but is yet to be duplicated.*

SUMMARY

Agriculture was America's central pursuit when this country was founded. Informal attempts at agricultural education were made by sharing information through agricultural societies and Farmers' Institutes. The Morrill Act created a college for the industrial classes to include the study of agriculture; the Hatch Act established the research component to strengthen the science of agriculture. However, it was the leadership of Seaman Knapp who established the demonstration method that became the model of a national network of county agents, now called the Cooperative Extension Service.

Time Line

1783 United States became a free nation

1785 Philadelphia Society for Promoting Agriculture established as first agricultural society

1790 United States Patents Office created

1823 Gardiner Lyceum established as first school for agriculture instruction

1855 Michigan established the first agriculture college

1862 Morrill Act of 1862 gave endowment for creation of land-grant schools

1862 Iowa became the first state to establish a land-grant college under the Morrill Act

1862 Organic Act created the United States Department of Agriculture

1863 First Farmer's Institute organized in Springfield, Massachusetts

1875 First American Experiment Station established in Middletown, Connecticut

1887 Hatch Act created experiment stations for land-grant colleges

1890 Morrill Act of 1890 led to establishment of black land-grant colleges

1891 Rutgers offers first agriculture extension programs

1896 Seaman Knapp began work as land developer in Louisiana

1899 Moveable school popularized by George Washington Carver at Tuskegee

1902 Walter Porter Community Demonstration Farm established by Seaman Knapp

1903 Field agents appointed by the USDA to work in southern states

1906 Thomas Campbell appointed first black USDA field agent

1906 W.C. Stallings appointed first county agent in Smith County, Texas

1907 First USDA work with boys established in Mississippi

1910 First USDA work with girls established in South Carolina

1910 First USDA home demonstration agents appointed

1912 First black home demonstrators appointed

1914 Smith-Lever Act created the Cooperative Extension Service

1916 Memorandum of Understanding signed

DISCUSSION QUESTIONS

1. Discuss the needs of education, agriculture, and science in America around the turn of the century. How did they interrelate? What influence did each have on the other?
2. What were some major events in history that hindered the land- grant movement? Helped the movement?
3. Why was the land-grant college act a revolutionary idea?
4 Explain the "broad" and "narrow" gauge philosophies of education? Is there any evidence of either philosophy in your college curriculum?
5. Who are some of the key individuals that promoted the land-grant colleges? The experiment stations? The Extension movement?
6. What were the key legislative acts that lead to the creation of the Cooperative Extension Service? Explain the major aspects of each act. Discuss early attempts at "Extension education."
7. Why was the "demonstration method" so successful?
8. How did the resistance of farmers contribute to work with youth and women? Would the development of this work have progressed without federal assistance?
9. If you could structure or create an Extension Service today, what would it be like?
10. Discuss how Extension may change in the next twenty years.

REFERENCES

Bailey, J. C. 1945. *Seaman A. Knapp: schoolmaster of American agriculture.* New York, NY: Columbia University Press.

Bliss, R. K., et al. 1952. *The spirit and philosophy of Extension work.* Washington, D.C.: USDA Graduate School and Epsilon Sigma Phi, National Honorary Extension Fraternity.

Boone, E. 1989. Philosophical foundation for Extension. In D. J. Blackburn ed. *Foundations and changing practices in Extension.* Guelph, ONT., Canada: University of Guelph.

Cline, R. 1936. *The life and work of Seaman A. Knapp.* Nashville, TN: George Peabody College for Teachers.

Eddy, E. D., Jr. 1957. *Colleges for our land and time: the land-grant idea in American education.* New York, NY: Harper and Brothers.

Elliott, L. 1966. *George Washington Carver: the man who overcame.* Englewood Cliffs, NJ: Prentice-Hall.

Evans, J. A. 1938. *Recollections of Extension history.* Raleigh, NC: North Carolina State College of Agriculture and Engineering of the University of North Carolina and the USDA, cooperating.

Extension Committee on Organization and Policy. 1981. *Report of the National Task Force on Extension Accountability and Evaluation System.* Washington, D.C.: Extension Committee on Organization and Policy, Extension Service-USDA.

Extension Committee on Organization and Policy. 1990. *Strategic directions of the Cooperative Extension System.* Extension Service, USDA: AD-BU-5584-S. Minneapolis, MN: University of Minnesota.

Fiske, E. P. 1989. From rolling stones to cornerstones: anchoring land-grant education in the counties through the Smith-Lever Act of 1914. *Rural Sociologist* Fall: 7–14.

Joint USDA-NASULGC Committee on the Future of Cooperative Extension. 1983. *Extension in the '80s: a perspective for the future of the Cooperative Extension Service.* Madison, WI: University of Wisconsin-Extension.

Martin, O. B. 1941. *The demonstrative work.* San Antonio, TX: Naylor.

Ode, A. 1991. Whatever Happened to the County Agent? *American Horticulturist.* 70:4. Alexandria, VA: American Horticultural Society.

Prawl, W., Medlin, R., and Gross, J. 1984. *Adult and continuing education through Cooperative Extension Service.* Columbia, MO: University of Missouri Press.

Sanders, H.C., ed. 1966. *The Cooperative Extension Service.* Englewood Cliffs, NJ: Prentice-Hall.

Sanderson, D. R., ed. 1988. Understanding Cooperative Extension: our origin, our opportunities. *Working with our publics,* Module 1. Raleigh, NC: North Carolina State University.

Rasmussen, W. D. 1989. *Taking the University to the people. Seventy-five years of Cooperative Extension.* Ames, IA: Iowa State University Press.

Rasmussen, W. D., ed. 1975. *Agriculture in the United States: a documentary history.* Vol. 1. New York, NY: Random House.

Reeder, R. L., ed. 1979. *The people and the profession.* Washington, D.C.: Epsilon Sigma Phi.

True, A. G. 1929. *A history of agricultural extension in the United States, 1785-1925.* USDA Miscellaneous Publication No. 15. Washington, D.C.: U.S. Government Printing Office.

United States Department of Agriculture. 1993. *Agriculture statistics.* Washington, D.C.: U. S. Government Printing Office.

Vines, C. A., and, Anderson M. A., ed. 1976. *Heritage horizons: Extension's Commitment to people.* Madison, WS: Journal of Extension.

Wessel, T., and Wessel, M. 1982. *4-H: an American idea: 1900-1980.* Chevy Chase, MD: National 4-H Council.

Challenge and Change . . . A Blueprint for the Future. 1983. Washington, D.C.: Extension Service-USDA.

C H A P T E R

"The work of the individual still remains the spark that moves mankind forward."
—*IGOR SIKORSKY*

Organization, Structure, and Administration for Extension Programs

Key Concepts

- Organizational structure and funding
- Staffing in extension
- Staff development and training
- Assessing performance
- Legal considerations
- Administration of extension programs

ORGANIZATION AND STRUCTURE

The Cooperative extension System is a unique and sometimes complex entity, linking together education and research resources at the federal, state, and local levels. Several factors are key to understanding the cooperative structure necessary for such an entity to exist.

The federal partner is the United States Department of Agriculture (USDA). Nine separate units (divisions) report directly to the Deputy Secretary for Agriculture. A reorganization of the USDA in 1994 placed the

Cooperative Extension System in the sub-unit Cooperative State Research Education and Extension Service (CSREES) under the office of the Under Secretary for the Research, Education, and Economics Unit. Organizational charts can be found in Appendix 6.

The traditional organizational structure of cooperative extension is based on cooperation in funding and programming on three levels. Three partners in funding are the federal, state, and local governments. The three partners in programming are those entities previously mentioned: USDA, the state land-grant institutions, and the local county/parish offices. However, in some states, counties offices have been eliminated and others have not activated local structure for extension work (Hoffert, 1979). Two documents establish the unique terms of this cooperative effort. The Smith-Lever Act, previously discussed in Chapters 1 and 2, provided guidelines for establishing the Cooperative Extension Service as a part of each state land-grant university. In addition, the 1916 Memorandum of Understanding between USDA and the land-grant institutions outlined the basic features and function each partner provides. The Memorandum was discussed in greater detail in Chapter 2.

Land-grant colleges and universities have formed an alliance known as The National Association of State Universities and Land-Grant Colleges (NASULGC). This association provides the mechanisms for land-grant institutions and the USDA to work together in formulating programs and policies. Programs and policies of specific concern to extension are formed by the Extension Committee on Organization and Policy, commonly know as ECOP. ECOP's plans and policies are deliberately broad to ensure that each state and local program has the flexibility to address local issues and concerns. ECOP addresses all aspects of extension's organization as it relates to programs, finances, its relationship to USDA, research, resident instruction, and farm organizations.

Extension programming and funding also benefit from private support. Private support comes from individuals as well as organizations and corporations. Donations are made at the local, state, and national levels in all program areas. The 4-H awards and recognition program, in particular, benefits significantly by receiving most of its funds from private support sources.

Funding

Although the federal, state, and local governments each contribute to funding extension programs, the contributions are not in equal proportions. The Smith-Lever federal funds are distributed to the states by law for certain expenses and are called "formula funds." Additionally, there are funds for designated programs reviewed on an annual basis, called "soft money." Although percentages vary from situation to situation, in 1993, the average county extension program received more than a third of its funding from the state, less than a third from local sources, and less

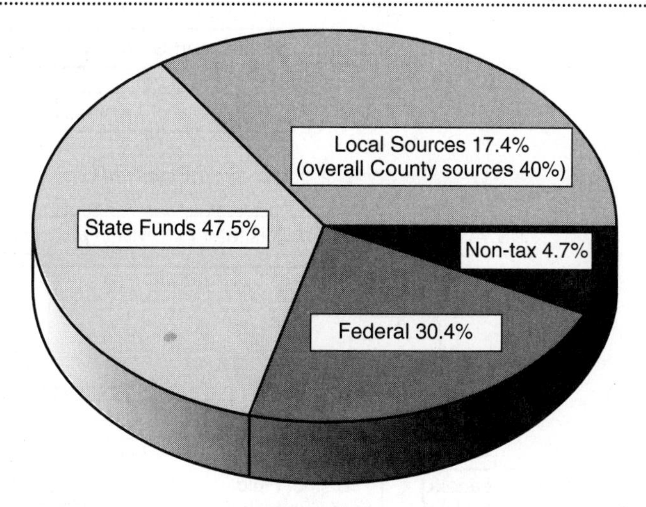

Figure 3–1 ***Extension County Offices Profile: Source of Funding in Counties. PDEES-USDA, County Office Study, 1992.***

than a third from federal and other sources, see figure 3–1. Funding of the total Extension system is primarily from federal income taxes, state taxes, and property taxes. This information is presented in figure 1–2 in Chapter 1.

At the local level, the persons charged with overseeing how money is spent are elected officials, county commissioners, a board of supervisors, or a county council. State funds are appropriated by state legislatures, and federal funds by the United States House of Representatives. State funds may be a part of the appropriation to the land-grant institution or a separate allocation from the state budget. As governmental funding for extension programs continue to decrease, many states are exploring alternative funding sources such as external grants, contracting out specific jobs (e.g., 4-H activities coordinator), and options and fees for programs and services.

Programming

Extension's tri-level partnership also extends to determining program direction. Each of the programming partners—USDA, land-grant universities, and county offices—have input in program decisions. As the federal partner, USDA establishes guidelines for programs, determines issues or initiatives of national scope, and provides program support. State specialists and administrators at land-grant institutions assist in determining statewide issues and initiatives, providing training for county professionals, and conducting and disseminating relevant research. A traditional organizational structure for state extension programs is illustrated in figure 3–2.

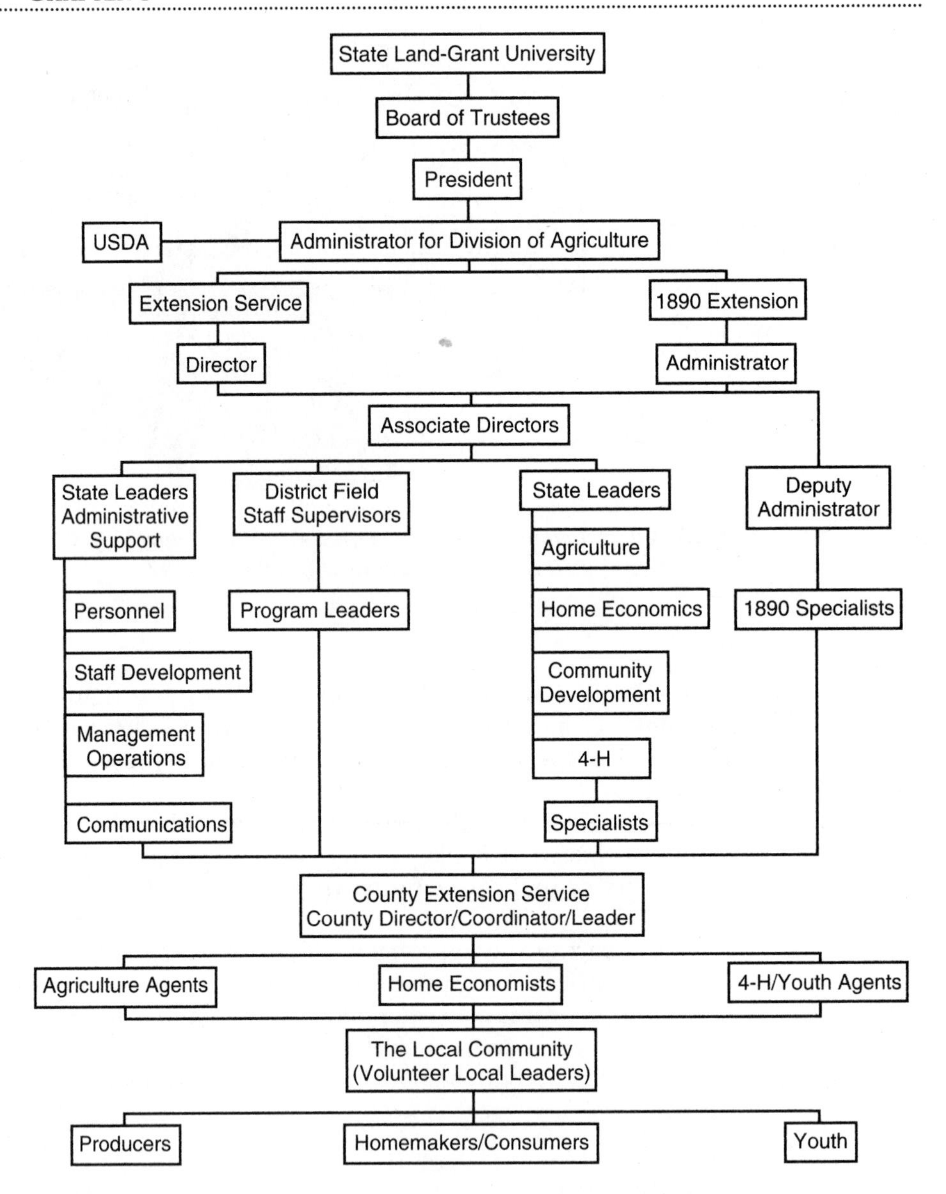

Figure 3–2 Hypothetical Organizational Chart of the Extension Service.

The basic operational unit of the partnership is the local city, county or parish, or multi-county program. Local units enjoy a great deal of freedom to plan, implement, and evaluate programs based on the needs of the local clientele. This unique educational network makes it possible to uphold the land-grant philosophy of making education available to all citizens by taking the university to the people. Development of extension programs will be addressed in detail in Chapter 5.

Structure

The local office is the traditional center for programming. County agents receive subject-matter support from state specialists; see diagram 1, figure 3–3. However, by the early 1990s, more than half of the states were experimenting with alternative organizational structures (PDE ES-USDA 1993). Nebraska was one of the first to replace single-county program units. There, eighty-seven single-county units were replaced by twenty-one program units (Rockwell, et al. 1993). In 1992, Iowa placed a local extension administrator in each county and instigated a system of field specialists who serve anywhere from a few counties to a quarter of the state. Illinois recently initiated clusters in some of its seven regions. A cluster may contain two or more counties and one or more education centers. Multi-county/state

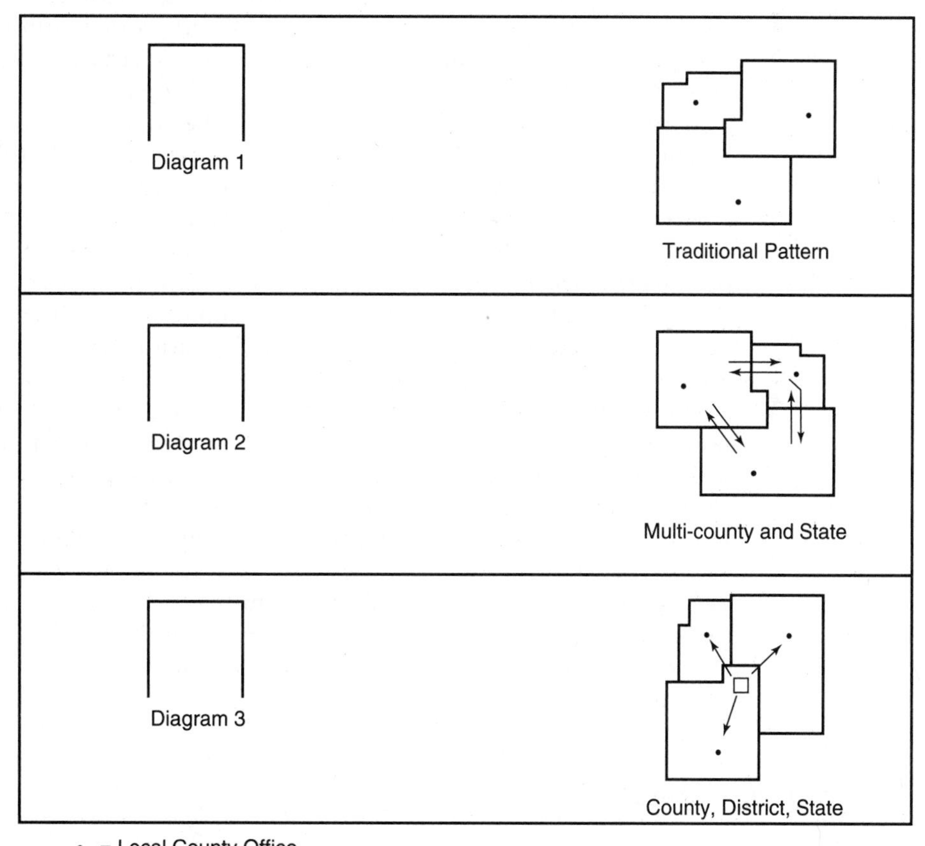

Figure 3–3 *Three Common Staffing Patterns. Adapted from Prawl, W.; Medlin, R.; and Gross, J. (1984).* **Adult and Continuing Education through the Cooperative Extension Service.** *University of Missouri Columbia.*

programming may take on different definitions. One example of multi-county combination involves an extension professional in a specific program area (e.g., 4-H or Home Economics) who assumes responsibility for all extension programs in that subject area for two or more counties. Another multi-county/state variation involves a county extension professional assuming the role of a field specialist for a particular subject area within a program area (e.g., Foods and Nutrition) and serving a multi-county area, with back-up support from state specialists. The direction of the arrows in diagram 2, figure 3–3 indicate the shared responsibilities between county units. Districts, areas, and regions have functioned for a long time, although primarily for administrative reasons. These groupings of counties within a geographic boundary usually have an assigned professional for administrative support. Some states also support program specialists at the district/region/area level who act as intermediaries between state and local programs. This staffing pattern involves county extension professionals working out of a specific county office. Their work is limited to work in that county with support from the district specialists who work the multi-county area. Both county and district staff are supported by state specialists. The arrows in diagram 3, figure 3–3 show the subject-matter support originating at the district office and spreading to the counties within the district.

Another organizational alternative is a university Extension system where the traditional agriculture and home economics content areas may be joined by other content areas in other colleges of the university, such as engineering, business, music, and design. Each college involved provides the specialists to reach clientele in their subject area. university Extension generally includes continuing and adult education areas as part of its focus. Some workshops and programs are offered off campus in a similar format to the Cooperative Extension System programs, other courses (both credit and noncredit), classes, and seminars are made available on campus. Many university credit courses offer continuing education units (CEUs) to practicing professionals to encourage them to remain current in their fields (e.g., nursing, teaching). The university Extension system is advantageous in that more of the research expertise is available for outreach to the citizens of the state. However, federal formula funds are available only to the Cooperative Extension Service and not to the university Extension system. Wisconsin, West Virginia, and Missouri were some of the first states to initiate a university Extension system.

STAFFING IN EXTENSION

Extension professionals, regardless of their location throughout the state are land-grant university employees. Their classification within the university structure varies from state to state. extension personnel in many states are considered faculty of the university and as such carry all the rights and responsibilities of any other full faculty member, including documenting performance and achieving professional stature through a formal process known as promotion and tenure. Some university systems use separate classification systems for on-campus academic faculty and CES faculty. The rational for the separate systems is that CES faculty conduct educational programs in a nonformal setting that is significantly different from a course of instruction completed in a university setting. Still in other university systems, Extension personnel are classified as professional staff.

Extension personnel can most generally be classified into one of five staffing roles: administration, program specialists, county agent, paraprofessional or program assistant, and support or clerical.

Administrators for the extension program are found at the state, district, and county/parish levels. Typical of most organizations, the higher on the hierarchy one is, the greater the concentration of administrators. Administrators at the state level assume such titles as vice president, chancellor, dean, associate dean, or director and have responsibility for all programs, personnel, and activities of the Cooperative Extension Service. The top administrator is supported by several others that compose the administrative team or "cabinet" for the organization. Each cabinet member generally assumes responsibility statewide for a major program area (agriculture, home economics, 4-H, etc.) or administration area (personnel, evaluation, operations, etc.).

Many states divide administrative and program responsibilities by region or district. A director (administrator) is employed for each region/district to work with the many counties/parishes in that district. The district/regional director is in constant communication with the counties/parishes in his or her area, and as such, this individual is frequently a part of the state administration team or cabinet. Depending on the state, district directors may be housed at the land-grant university and travel to their district on a regular basis, or have an office in the district they serve.

Since cooperative extension serves every community in the state, a designated individual at the local level is necessary to handle the daily operations and administrative duties. Most frequently referred to as the "county director," this individual usually shares both programming and administrative roles. The number of staff in local offices varies greatly. Sparsely populated areas may have only one extension agent, or share that agent with a neighboring community. Urban or metropolitan areas may have as many as thirty personnel on staff. In larger offices, one individual is usually designated a full-time administrator.

Program specialists are experts in a particular subject who are trained to translate and disseminate researched-based material. Specialists provide a vital link in the information-dissemination process. Specialists are considered experts in their topic area and use varied techniques to share new information and solve problems. Specialists and county staff are in constant communication with one another through telephone, electronic mail, and personal contacts. Specialists help county faculty plan, carry out, and evaluate educational programs. They write bulletins and newsletters used by both agents and clientele and they help answer clientele questions. In addition, they conduct training programs to update extension agents and serve as expert resources in local workshops sponsored by the county agent. Most specialists conduct research in their designated area. Specialists are supervised by a state-level administrator, but they must possess expertise not only in their topic area, but also in working with people of all ages and backgrounds. All state extension programs have specialists in a variety of areas. Some states also employ subject specialists at the district or regional level.

The county agent is the heart and soul of the Cooperative Extension Service. Many different staffing patterns are being implemented throughout the nation; however, the traditional pattern in which an agent works in one county is still the predominant approach. This professional staff person working at the local level has constant contact with the clientele he or she serves. The county agent, sometimes known as extension educator, change agent, teacher, or social activist, serves a unique role in the community and is certainly the most visible of all extension staff. The county agent serves as an educational broker for the community. Agents are employed as agriculture, home-economics, 4-H/youth, or community-development generalists in most states; however there are also positions with specialized work assignments. Some examples include positions in water quality, food and nutrition, and horticulture. The agent provides leadership and expertise in utilizing available resources to extend knowledge and solve problems. The county agent reports directly to a district- or state-level administrator.

In some local programs, additional help is sought by hiring a paraprofessional or program assistant. The paraprofessional is supervised by and serves as an assistant to the county agent and usually lives in the community in which he or she works. Paraprofessionals often are not required to have a college degree; they most frequently assist with coordinating and organizing many of the various activities and events that the office sponsors. Educational programming itself is coordinated by the county agent.

Support staff have a vital function in any organization and are found at every level of the organization. Receptionists, secretaries, bookkeepers, and office managers are frequently the first contact clientele have with the organization. Sometimes lasting impressions are made during that initial contact. Smaller offices tend to have only one support staff member who assumes all support duties. Although this individual may assist all other staff in the office, direct supervision is usually the responsibility of the county director. Offices with additional support staff will usually designate one individual as the office manager. It is the office manager who provides daily supervision of the other support staff. The office manager reports directly to the county director. Like other extension employees, support staff are employees of the land-grant university system. However, because of the partnership between state and local governments, most support staff employees observe the same classifications, policies, and procedures of other county government employees in that community.

A distinctive classification of "employee" essential to successful Cooperative extension programming includes the millions of volunteer staff that share their time and talent each year. Volunteers are active in all dimensions of extension efforts, including serving on advisory committees, teaching programs and projects, and leading 4-H club activities. Since the beginning of the Extension Service, agents have relied on volunteers to extend the services provided. Working with volunteer staff poses some distinct differences from working with paid employees. Some of the differences and the similarities will be explored in Chapter 9.

EMPLOYMENT IN THE COOPERATIVE EXTENSION SERVICE

As the Extension Service evolved and expanded, the roles and responsibilities of the extension professional also changed. Staffing considerations and finding the right person to meet changing demands became more important than ever.

In 1993, the Personnel and Organization Committee of the Extension Committee on Organization and Policy identified sixteen core competency areas that all extension professionals should possess. The core competency areas include:

1. Applied Research
2. Change Management
3. Communications and Human Relations
4. Computer Operations and Software
5. Conflict Resolution
6. Cooperative Extension System
7. Educational Programming (Program Development)
8. Evaluation and Accountability
9. Instructional Development and Learning
10. Marketing and Public Relations
11. Organizational Development
12. Personal Organization and Management
13. Professional and Career Development
14. Public Policy Education ("Citizen Politics")
15. Resource Development and Management (Human and Financial)
16. Strategic Planning

In addition, Hahn (1979) identified clusters of skills, abilities, and other characteristics essential to all successful extension professionals.

Although these characteristics are broad enough in nature to apply to many professional positions, extension's commitment to research and public education of clients of all ages and backgrounds makes these particular characteristics especially relevant. Hahn clusters the characteristics into seven groups: commitment to the job, communication skills, interpersonal skills, positive attitude, program development and direction, problem solving, and self-confidence, see figure 3–4. The interlocking circles demonstrate that many of the skills or characteristics are interdependent.

These qualifications are universal of all extension professionals regardless of the position within the organization. Most position announcements clearly identify these as desired qualifications for employment.

A career with the Extension Service requires a unique and complex individual. Each day is a new and varied experience. A degree in an appropriate area of study is necessary for any extension position. However, the degree alone is not the only qualification looked at during the hiring process. Enjoying working with people of all ages and backgrounds is also an essential requirement. Technical knowledge and expertise are important, but ineffective without the ability to relate to the clientele being served. A significant portion of an extension professional's time is spent interacting with people, problem solving, and managing conflict situations.

The Extension Service is an exciting career choice, but it is also a demanding one. Long hours, evening and weekend meetings, time away from family, and juggling multiple requests can be frustrating, but the rewards that come from working with diverse audiences, job variety, and knowing that you may have made a difference in someone's life are high. Planning, organization, and time-management skills are essential to ensure a positive experience. Burnout and frustration are two of the most frequent reasons cited for terminating employment with the Cooperative Extension Service. Time-management and organization skills will not eliminate the many demands an extension professional faces, but they can give the individual greater control in carrying out personal and professional responsibilities.

TRAINING AND ORIENTATION

When the Extension Service first began, early workers were required to have only a certificate or degree from a four-year agricultural college or its equivalent. Since those early days many changes have occurred that firmly established extension education as a career option. As a result, extension began to look seriously at training needs for both future and current staff.

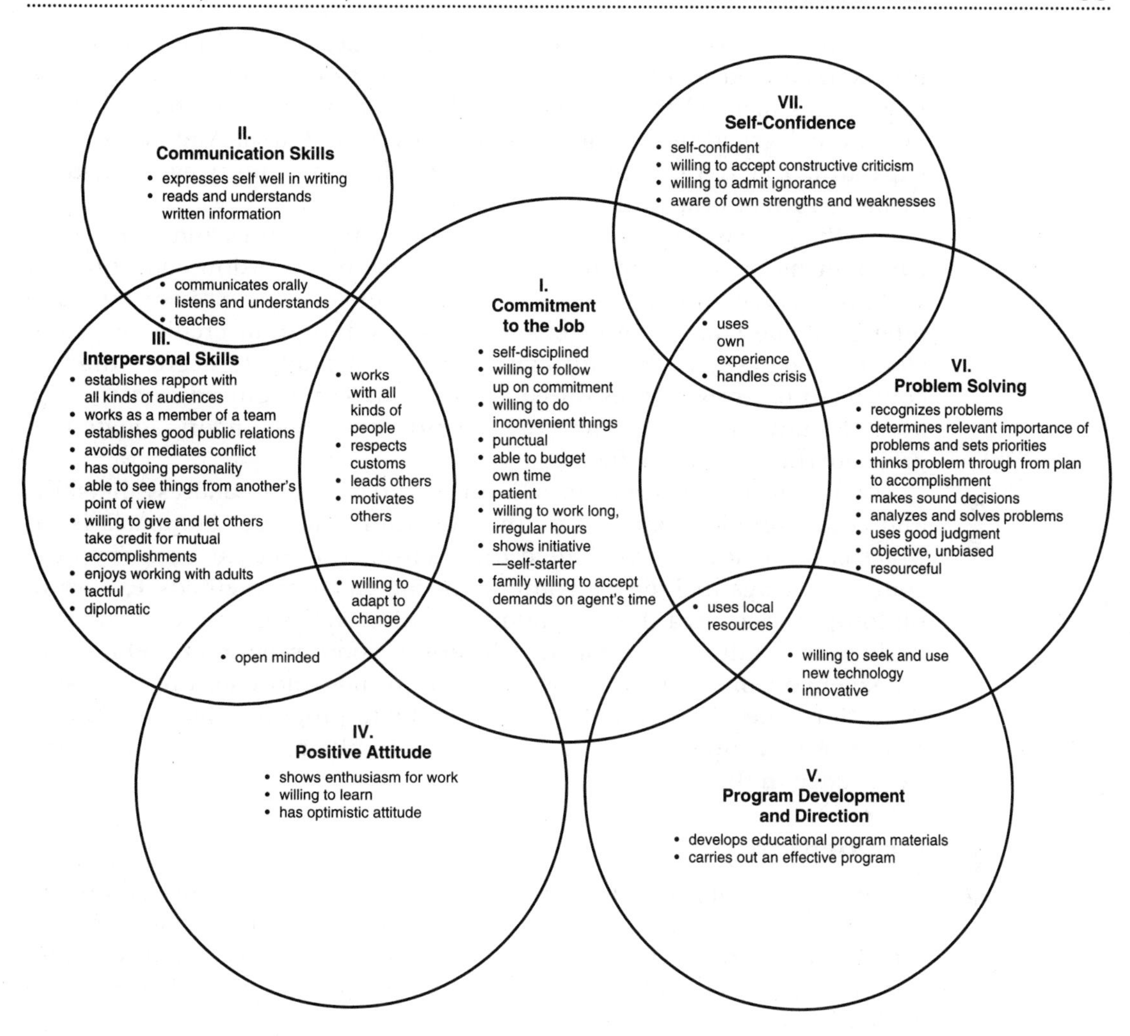

Figure 3–4 *Clusters of Skills, Abilities, and Other Charcteristics. Hahn, C.P. (May 1979) Summary Report:* **Development of Performance Evaluation and Selection Procedures for Cooperative Extension Service.** *Washington, D.C. USDA.*

Preservice Training

Today the minimum requirement for employment in the Cooperative Extension Service at the local level is a bachelor's degree from an accredited institution. However, the majority of state programs require that the individual also possess a graduate degree at the master's level. Since extension's roots and traditions are firmly founded in agriculture and family, it is not surprising that most states hiring extension professionals looked for individuals with degrees in agriculture or home economics. But as perceptions of extension professionals broadened and the agent began to be viewed more and more as an educator, position announcements expanded to include degrees in education or related areas. Still, most states insisted that an individual possess at least one technical degree in an agriculture- or home-economics-related area. And, despite the expanding role of the extension professional, many county agents had little or no formal preparation as educators.

As the Extension Service evolved, several universities and colleges began offering bachelor's and graduate degrees in extension education. In 1929, the University of Wisconsin, became one of the first institutions to offer graduate courses in extension methods and grant credit for extension-related research that ultimately lead to a graduate degree. Interest continued to grow; extension education now is recognized as a specific area of study and today numerous institutions offer undergraduate and graduate training in the field.

Course work in extension education is designed to address the skills and competencies required of an extension educator. Emphasis is on practical application grounded on a sound theoretical base. Typical course offerings include program planning, teaching methods, extension philosophy and organization, administration and supervision, program evaluation, adult and nonformal education, youth program development, management of volunteer programs, and trends and issues in extension education. Many undergraduate and graduate programs also provide internships and supervised field experiences to allow the student first-hand experience in the profession.

Orientation

When an extension professional is hired, most states provide a thorough orientation. Although methods used to conduct new staff orientation vary greatly, most programs offer information on cooperative extension philosophy, structure, and policies and program development, delivery, and evaluation.

Staff Development

In many states, new extension employees begin their career by developing a professional development plan. Personal and professional development through ongoing training efforts should be a goal of every extension employee, regardless of whether or not it is a function of job orientation. Technology has enabled anyone to access new information, innovations, or the latest research that in the past would normally have been provided by the agent. Without frequent updates and training, any employee can become outdated and unable to provide the recommendations for current practices. Personal and professional development comes in many different forms. Professional development is supported by the Extension Service for its employees through in-service training and involvement in professional organizations.

In-service Training. In-service training is a widely used method to provide training in both subject-matter areas and methodology. In-service training programs, usually coordinated at the state level, provide the opportunity for employees to receive training in the most current issues and methods without taking a leave from their job. In-service is usually an intensive educational effort lasting from one day to one week. Many state extension programs require employees to participate in a set number of in-service training days each year.

Professional Organizations. Membership in professional organizations is also an excellent professional development tool. In addition to the many professional organizations and societies related to special interests, there are several organizations specifically for extension employees. Some extension organizations were developed around areas of primary responsibility, others evolved because of special needs or interests. Three organizations were specifically formed for employees that worked in a primary area of program responsibility. The National Association of County Agriculture Agents (NACAA), the National Association of Extension Home Economists (NAEHE), and the National Association of Extension 4-H Agents (NAE4-HA) each sponsor a national conference. Benefits of membership include opportunities to attend workshops, seminars, research meetings, and field trips, to network with colleagues of similar interests and to take part in organizational leadership opportunities, and subscriptions to professional publications and newsletters. Each state also has a state chapter of each of these organizations.

Other professional organizations directly associated with the Cooperative Extension Service include the Association for International Agricultural and Extension Education (AIAEE) and the National Extension Honorary Society, Epsilon Sigma Phi.

Personal Reading. Every extension professional should make reading professional literature a priority. New information, technology, and research is flooding the market. Making the time to keep abreast of the most current information can make a considerable difference in how credible an extension professional is perceived to be by clients. In addition to technical journals related to specialized areas of interest, extension educators should read the *Journal of Extension*. This official publication of the extension profession provides feature articles on the current issues and research, a futures perspective, research findings, and hints and ideas. Utilizing modern technology, any employee with a computer and a phone line can access the quarterly publication free at home or in the office.

Computer networks. Networks are easy to access and provide the latest information worldwide. Personal communications, newsletters, and technical reports have almost become obsolete. Through the use of computer networks, information from anywhere in the world is easily available at a moment's notice.

Mentoring programs. Mentors are an additional way for a new employee to make a smooth transition into the profession. Mentoring can occur at any level of the organization. A mentor is a guide. An experienced employee is selected as the mentor. The new employee (mentee) calls upon the experience and wisdom of the mentor in new situations. As the mentee's competence and confidence increase, the role of the mentor slowly diminishes.

ASSESSMENT OF PERFORMANCE

On a regular basis, all extension employees are required to be evaluated on their job performance. Although a variety of procedures and methods exist to complete the evaluation process, most organizations require the immediate supervisor to conduct the appraisal, complete the required paperwork, and inform the employee of the results. The individual conducting the appraisal should be in a position to regularly observe employee performance.

Assessing employee performance satisfies several functions within the organizational structure. In the Cooperative Extension Service, performance appraisals are most commonly used to let employees know where they stand in relation to performance objectives and organizational expectations. Employee appraisal should be based on the job description. An analysis of the job should identify specific job behaviors and performance outcomes that are measurable.

Feedback of performance outcomes is particularly important for designing personal training and development plans in areas identified for growth or improvement. Recognition of high performance can be a personal motivator by fostering feelings of achievement and accomplishment. Additionally, appraisal systems are utilized to make administrative decisions about employees. Administrative decisions linked to performance include items such as promotions, transfers, and salaries.

The organization should clearly identify the purpose of and describe the procedures utilized in any performance appraisal system in a policy and procedures manual. If no manual is in existence, the procedures and policies should be clearly communicated to employees at the time of hire and this communication should be documented.

Two specific performance measurement issues need to be addressed further. As already discussed, measurement of performance will be valid only if it is based on the specific dimensions of the job's activities. Performance of job activities are most frequently approached in relation to results of the job, or behaviors that lead to those results. Most performance measurements are multi-dimensional in that they address both outcomes and behaviors. The second issue involves identification of performance standards: what constitutes a good, bad, or neutral rating. It is imperative, from a legal perspective, that these issues be resolved at the institutional level prior to any performance assessment.

Although there are multiple ways to conduct an assessment of performance, the performance appraisal form, or instrument completed by the supervisor is the most common. A feedback conference, reviewing the instrument, results, and recommendations, should be scheduled with the employee.

LEGAL ASPECTS OF EMPLOYMENT AND PROGRAMMING

Since the early 1960s a variety of legislation has been passed that has significantly influenced hiring practices, not only in cooperative extension , but throughout all institutions in the nation. Legislation concerning fair employment, affirmative action, and Americans with disabilities has impacted both employment and programming efforts in the Cooperative Extension Service.

Title VII of the Civil Rights Act of 1964 established the Equal Employment Opportunity Commission (EEOC). In addition to enforcing other legislation, the EEOC is responsible for prohibiting employment discrimination based on race, color, national origin, sex, age, and physical handicap; enforcing equal pay for equal work; and coordinating federal equal employment opportunity programs. An amendment to the legislation in 1972 included the Extension Service in all boundaries of the law. EEOC legislation has impacted Extension through employment practices and programming efforts. All extension printed materials (i.e., publications and correspondence) contain some variation of a nondiscrimination statement. The nondiscrimination statement of the USDA is as follows:

> *The United States Department of Agriculture (USDA) prohibits discrimination in its programs based on race, color, national origin, sex, religion, age, disability, political beliefs, and marital or familial status. (Not all prohibited bases apply to all programs.) Persons with disabilities who require alternative means of communication or program information (braille, large print, audiotape, etc.) should contact the USDA Office of Communications.*

Fair employment legislation was implemented to eliminate discrimination in the workplace by enforcing policies that ensure that hiring and promotion decisions are determined based on job-specific criteria and not on personal characteristics such as gender, age, race, religion, color, national origin, or physical handicap. Discrimination can be classified as disparate treatment, disparate impact, or sexual harassment. Disparate treatment occurs when an individual protected under one of the above categories is treated differently than another employee or applicant. An example would be a man and a woman receiving different sanctions for the same rule violation. Disparate impact occurs when factors that appear to be neutral actually serve as barriers, such as by eliminating an unusually high number of women or minority applicants. Sexual harassment constitutes any action of a sexual nature that results in an intimidating, hostile, or offensive work environment. An example of harassment is for a supervisor to demand sexual favors in return for fair treatment. Inappropriate language, jokes, photos, and stories as well as unwelcome or unsolicited touching or looks are also considered harassment. Although more women report incidents of sexual harassment, men can also be victims.

The primary purpose of affirmative action programs is to achieve parity between the percentage of women and minorities reported in the labor market and the organization doing the hiring—the Extension Service. Affirmative action plans assist in achieving equitable representation in the work force. Although affirmative action is sometimes misinterpreted as allowing preferential treatment, this practice would be in strict violation of the Civil Rights Act. When developing affirmative action plans, organizations should consider areas of underutilization or concentration. Underutilization refers to low involvement in certain positions held by women or minorities: for example the small percentage of women nationally in agricultural agent or specialist positions could constitute underutilization. Concentration, on the other hand, is in effect when large numbers of women or minorities are found to be in low-paying positions.

The most recent legislation that has had an impact on the Cooperative Extension Service and other federal programs is the 1994 Americans with Disabilities Act (ADA). The legislation further extends previous efforts by providing additional protection against discrimination for those with disabilities. The law's implications for program delivery and management are significant. Programs must be held in locations and facilities that are accessible to all individuals. Also, for example, a hearing-impaired individual may request that an individual be provided to present a program in sign language.

The same legislation that governs many employment practices also applies to programming efforts. All programs and activities sponsored by the Cooperative Extension Service must be open to any interested person regardless of race, religion, gender, national origin, age, color, or physical disability. All correspondence and literature developed through the Cooperative Extension Service includes some statement of intent. To ensure compliance, county and state programs must submit annual reports to the USDA. Additionally, programs may be reviewed by an interview team on a periodic basis. Since participation in extension programming is voluntary, complete compliance can be difficult. However, the Extension Service must be able to demonstrate that "all reasonable effort" was made to promote and encourage participation from all protected groups.

ADMINISTRATION OF EXTENSION PROGRAMS

Educator, problem-solver, change agent, counselor, planner, confidant, supervisor, activity coordinator, organizer, media contact, consultant, and leader are just a few of the varied roles the extension professional assumes. The extension professional holds a unique position within the community. Known and respected not only for technical expertise but also for the daily involvement in planning programs, organizing meetings and activities, supervising volunteer and paid staff, responding to clientele needs, meeting with opinion leaders and policy makers, and assessing program impact, the extension professional must be adept in all aspects of management and leadership. Skills in planning and organization are critical to successfully meet the many program demands. The ability to work with people of all ages and backgrounds is essential. As a change agent and problem solver, the extension professional is viewed and respected as an opinion leader, facilitator, and link to resources.

Theory into Practice

Effective leaders recognize that many factors influence the selection of an appropriate leadership style for any given situation. Because an extension professional assumes so many different roles, flexibility, and the ability to assess situations are necessary skills. In any given day, the extension agent may need to address specific clientele concerns, meet with elected officials, teach a workshop, chaperon youth programs, and meet with an advisory committee, board, or council. Each task, each group, and each situation may require a different approach or leadership style. For example, when working with an advisory committee, a participatory style may be desired to encourage growth and involvement of the members as well as accomplish a specified task; however, when focusing on a very task-oriented project with a short time line a directive style may prove more efficient and effective.

Although there are many situations when a directive leadership style is appropriate in extension, the work of extension—"helping people to help themselves"—encourages a participatory approach. The grassroots philosophy supports participation, involvement, and empowerment of the people. Based on more participatory styles of leadership, Kouzes and Posner (1987) have identified five leadership practices of exemplary leaders. The very nature of the profession requires that extension professionals demonstrate each of these practices on a daily basis. According to Kouzes and Posner, exemplary leaders:

1. **Challenge the process** by challenging the status quo, seeking challenges, taking risks, and adapting to change

   (As change agents the focus of extension professionals is to promote and facilitate change—new ideas and innovations are diffused every day to improve the quality of life.)

2. **Inspire a shared vision** by looking ahead and anticipating, by explaining the "why" before the "how." They become excited and this becomes contagious. They set goals with the group and point them in the right direction.

   (Although extension assists individuals with problem solving, through education and planning the goal is to be proactive—to anticipate needed change and address it before it becomes a problem. For example, a program on water quality can help maintain safe standards before cleanup, health, and safety become an issue.)

3. **Enable others to act** by involving them in decision making, inspiring trust and mutual respect within the group. They transfer power to others by getting others to own projects, and by managing conflict.

   (Extension clientele are involved in the program development process in many different ways and are actively engaged in finding solutions to their own needs and problems. Involvement in decision and policy making is also evident through clientele involvement in advisory and program committees, volunteer commitments, boards, and support councils).

4. **Model the way** by setting examples and getting individuals to feel good about themselves without needing to prove themselves to others. They model positive self-concept and are able to break large projects into smaller chunks.

   (Extension professionals are an integral and respected part of the communities they serve. They are viewed as both role models and opinion leaders).

5. **Encourage the heart** by recognizing contributions, giving praise, celebrating accomplishments, and sharing the news of the group's successes.

   (The extension professional is visible at many community events and activities and at most extension functions. Providing support, monitoring a client's progress, promoting programs and activities through the media, recognizing achievements, and celebrating successful efforts are just a few ways of being encouraging).

MANAGEMENT, LEADERSHIP, AND SUPERVISION

Understanding principles of management and leadership theories is just a beginning to translating management theory into practice. The market today is flooded with books, articles, and journals on leadership, management, and supervision filled with prescriptions and philosophies on how to be the biggest and best, have happy employees, and ensure that you are not only doing things right but also doing the right thing. **Management** has been described as an art, a science, a body of knowledge, a process, and a set of functions. Despite this variety of descriptors, the common denominator found among most management definitions is this: effective utilization of available resources (human and material) to achieve common goals and objectives. Therefore it is the specific task of the manager to establish an environment that facilitates the attainment of the predetermined objectives. Scientific management led the way for the establishment of labor unions and an increased awareness of human resource management. Recent studies in management have concentrated on "administrative management," or how to be a successful executive, and "human relations theory" emphasizing individual behavioral patterns.

Buford (1988) shows that the work of any manager can be divided into five basic management functions: planning, organizing, staffing and human relations, leading and influencing, and controlling. Figure 3-5 illustrates the functions of the management process.

Supervision focuses more on influencing or directing the employee and less on the product itself. Effective supervision, however, should not

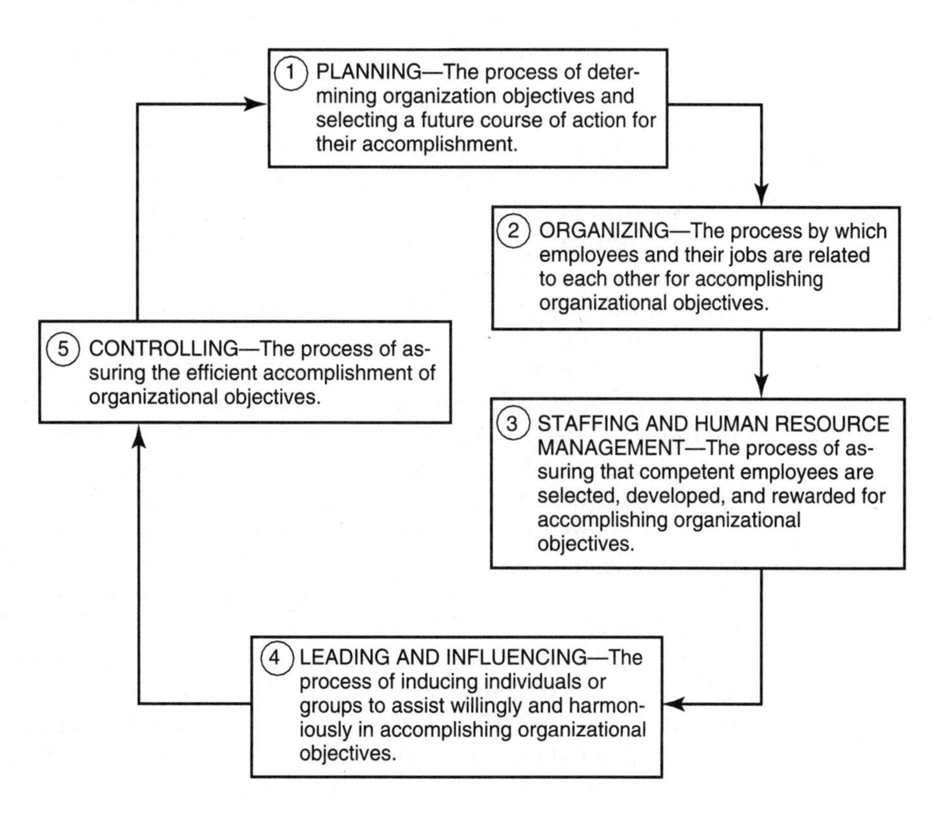

Figure 3-5 ***The Management Process. Buford, James. (1988) Management in Extension. Auburn, AL. Auburn University, Alabama Cooperative Extension Service.***

only recognize the employee as an individual; it should ultimately promote higher product quality and outcome also. Van Dersal (1974) identified the following six principles of supervision that can be applied to any situation involving a supervisor and subordinate.

- People must always clearly understand what is expected of them.
- People must have guidance in doing their work.
- Good work should be recognized.
- Poor work deserves constructive criticism.
- People should be encouraged to improve themselves.
- People should have an opportunity to show they can accept greater responsibility.

Since World War II, leadership has emerged as a recognized area of study and it continues to be a major focus in our rapidly changing society. Almost as many definitions of leadership exist as there are leaders. Clark and Clark (1994) define leadership as

> *an activity or set of activities, observable to others, that occurs in a group, organization, or institution involving a leader and followers who willingly subscribe to common purposes and work together to achieve them.*

Ideal leaders have been described as "transformational." The transformational leader is charismatic and inspiring, one who provides individualized consideration and intellectual stimulation. Yukl's definition of leadership emphasizes the influence the leader has (Clark and Clark 1994). Yukl says, "leadership involves influencing task objectives and strategies, influencing commitment and compliance in task behavior to achieve the objectives, influencing group maintenance, and influencing the culture of the organization."

In an exhaustive review of leadership studies, Clark and Clark (1994, p. 22) found some commonalities in leadership definitions. They found that in almost all definitions, leadership involves "leaders, followers, members, subordinates, or constituents as they interact, create visions, become inspired, find meaning in their work and lives, and gain in trust and respect." Leadership is not merely a function of position. There are those who are elected, appointed, or simply born into a position of power or authority. But holding a position of authority does not automatically make one a leader.

The early studies of leadership focused on two dimensions: traits and behaviors. Trait theory centered on the specific traits that make an individual a leader. A trait is defined as a distinguishing characteristic or quality. Trait theory held that a person was born either with or without the traits necessary to be a leader. As support for this theory diminished, researchers sought to describe those traits that separated seemingly natural leaders from all other persons. It was believed that through understanding these traits, leaders could be trained and the traits developed.

Behavioral leadership studies believed leaders should be characterized by the behaviors they exhibit and not the traits they possess. Behavioral leadership sought to explain what leaders do that causes others to follow them and to identify how they do it. The influence of the Ohio State University Leadership Studies conducted in the 1940s is still strong today. These studies identified two major dimensions of leadership behavior: consideration and initiation of structure. Later these two dimensions were more commonly described as relationship orientation and task orientation.

Consideration or relationship orientation is focused on the people doing the work. Emphasis is on the interaction between the leader and those doing the work. Initiation of structure or task orientation is centered on the work (task) to be accomplished. Emphasis is on meeting deadlines, meeting goals, and maintaining quality. Many studies since have sought to describe and categorize leader behaviors and to understand how those behaviors are developed or learned.

Behavioral studies were not without their limitations and by the 1960s attention was shifting to understanding how leadership behaviors vary depending upon the situational circumstances. The situational approach to leadership, often referred to as contingency theories, produced many studies that are still widely recognized and subscribed to today. One of the most common of these theories is Hershey and Blanchard's Situational Leadership Model.

Hershey and Blanchard's Situational Leadership Model is based on the relationship between the leader's directive and supportive behavior, and the amount of competence and commitment exhibited by the follower. Directive behavior is defined as the extent to which the leader engages in one-way communication, clearly defines and communicates to the follower his or her role, and supervises the follower in the process. Supportive behavior includes involvement in two-way communication, listening, providing support and encouragement, facilitating group interaction, and involving the follower in the decision-making process. When directive and supportive behaviors are plotted on a two-by-two matrix with a continuum of low to high, four leadership styles emerge. The four leadership styles are:

Directing High directive/low supportive. Leader provides specific instructions for follower and closely supervises task accomplishment.

Coaching High directive/high supportive. Leader explains decisions and solicits suggestions from followers, but continues to direct task accomplishment.

Supporting High supportive/low directive. Leader makes decisions together with the followers and supports efforts toward task accomplishment.

Delegating Low supportive/low directive. Leaders turns over decisions and responsibilities for implementation to followers.

Leadership and leadership development continue to remain a major focus of study and research. Apps (1993, p. 3) states "the next age will require new solutions, based on new ways of thinking. And, the new ways of thinking will also serve as foundation for approaches to leadership." Universities, government agencies and not-for-profit programs such as the Cooperative Extension Service are looking toward business and industry for models of new solutions to new problems. Total Quality Management (TQM) is a concept that originated in business and industry. Although it is most heavily utilized in Japanese corporations, it has gained popularity not only in American industry, but also in the public and government sectors. TQM is a comprehensive approach to management that produces products or services that meet or exceed the expectations of the organizations' customers. In its quest for quality, TQM centers on investigating the way in which work gets done from a systematic, integrated, and organizational perspective. An integral component of TQM is the utilization of work teams involved in all functions and personnel at all levels of the organization. Shared leadership and shared decision making are functions of the participatory leadership styles seen more and more frequently.

SUMMARY

Effective management of any organization requires awareness of the functions of planning, organizing, staffing, leading, and controlling. The county agent, whose primary function is engaging in educational efforts with clientele is the cornerstone of extension education. As needs and resources continually change, CES has reorganized to meet those changes in programs and in management. New staffing models have developed to maximize available resources. Training and development incorporates the latest and best technologies and methods, as well as the newest information and research. Changes in legislation has required CES to examine hiring and supervision practices to ensure not only compliance, but a positive and safe work environment. At every level of extension, changes in programming, administrative staff, and functions are necessary to ensure smooth and successful programming.

DISCUSSION QUESTIONS

1. What extension staffing patterns exist in your state? What are the pros and cons of changing this pattern?
2. What is an example of a management situation versus a leadership situation?
3. What characteristics should the ideal leader possess? How would these characteristics be assessed?
4. How should staff development and training programs be determined? Should the emphasis be on process skills or technical updates?

5. Attendance at staff training programs usually requires time away from the home county. Should a set number of staff training days be required for each employee?
6. Many extension agents are required to have a degree in a technical subject area, but training in education is optional. Should preservice or in-service training in educational philosophy and methods be required?
7. If you were the supervisor, how would you rate your employees?
8. Affirmative action has been accused of being a political "quota system." Respond to this accusation.

REFERENCES

Apps, J. W. Summer 1993. Leadership for the next age. *Journal of Extension* Vol. 31:3-5.

Buford, J. A., and Bedian, A. G. 1988. *Management in Extension.* 2nd ed. Auburn, AL: Auburn University, Alabama Cooperative Extension Service.

Clark, K.E., and Clark, M.B. 1994. *Choosing to lead.* Charlotte, NC: Center for Creative Leadership, Iron Gate Press.

Extension Committee on Organization and Policy. (August, 1990) *Conceptual framework for cooperative extension programming.* Washington, DC: Strategic Planning Council, Extension Committee on Organization and Policy and Extension Service.

Gordon, T. 1977. L.E.T.: *leader effectiveness training.* New York, NY: Bantam Books.

Hahn, C. P. May 1979. *Development of performance evaluation and selection procedures for the Cooperative Extension Service.* Summary Report. Contract Number 12-05-3-372. Washington, D.C.: United States Department of Agriculture.

Heneman, H.G.; Schwab, D.P.; Fossum, J.A.; and Dyer, L.D. 1989. 4th. ed. *Personnel/ human resource management.* Homewood, IL: Richard D. Irwin.

Hoffert, R.W. 1979. *The American states and the Cooperative Extension Service.* Ft. Collins, CO: Colorado State University, Cooperative Extension Service.

Kouzes, J. M., and Posner, B. Z. 1987. *The leadership challenge.* San Francisco, CA: Jossey-Bass.

PDE ES-USDA. (Oct. 28, 1993). County Office Study Planning, Development, and Evaluation, Extension Service. Washington, D.C.: United States Department of Agriculture.

Rockwell, S. K., et al. Fall 1993. From single to multi-county programming units: reactions to restructuring Extension in Nebraska. *Journal of Extension* 31(3): 17–18.

Sanders, H. C. 1966. *The Cooperative Extension Service.* Englewood Cliffs, NJ: Prentice-Hall.

Van Dersal, W. 1974. *The Successful Supervisor.* 3rd ed. New York, NY: Harper & Row.

C H A P T E R

"Education makes a people easy to lead, but difficult to drive; easy to govern, but impossible to enslave."
—*Henry Peter Brougham*

Program Areas in Cooperative Extension

Key Concepts

- Agriculture programs
- Home economics programs
- 4-H/youth programs
- Community development programs
- Other Extension programs

Traditional Program Areas

The four traditional program areas in Extension are agriculture, home economics, 4-H/youth, and community development. They provide educational programs for agricultural producers, families, school-age children, and communities. States differ in their titles for the program areas and may have additional program areas plus interdisciplinary programs. All program areas provide practical information and help people make better decisions. All are research based.

States also vary in the amounts of funding and staff time they devote to the various program areas. A major trend is toward issue programming supported by interdisciplinary work that cuts across traditional program areas. Many states involve a wide range of university resources in addition to the traditional agriculture and home economics research base.

AGRICULTURE

The agriculture program area includes the production, processing, marketing, and consumption of food and fiber. Agriculture has been dominant in terms of expenditures and personnel since the beginning of Cooperative Extension in 1914. Agriculture's dollar share of the budget for salaries is close to 50 percent. The amount spent for salaries is an accurate indication of the total amount spent for each area, because salaries make up almost 90 percent of the money spent for programming.

In the early years, the agriculture program area emphasized production and played an important role in sharing research about new varieties and methods with producers. In more recent years, emphasis has been on profitability and environmentally sound practices. Although everyone wants a food and fiber supply that is safe, plentiful, and affordable, increasingly consumers are concerned about agriculture's effect on the environment. The Extension system is listening to those concerns.

To be adopted today, agricultural practices must contribute toward a safe and abundant food supply while protecting the quality of the environment. Of particular concern are the safety of drinking water, the quality of irrigation water, and the safety of food grown in soil to which agricultural chemicals such as pesticides have been applied (Cooperative State Research Service, USDA 1993, p.10).

Goal

The main goal of the agriculture program area is to help producers earn a fair return on their efforts in an environmentally and socially responsible manner. Achieving this goal benefits not only producers but society as a whole as it promotes a stable and affordable supply of food and fiber.

The term "producers" includes farmers, ranchers, nursery workers, foresters, fishermen, and anyone who grows or raises a product. Horse and dog fanciers and plant lovers are included in this category, plus all who raise something just for their own enjoyment. In addition to producers, key players in the agricultural community are the input industries that service producers and the output industries that help move the product to the consumer, see figure 4–1.

Agriculture is the nation's largest employer (United States Department of Agriculture 1990), with about 18 percent of the domestic labor force employed in agriculturally related occupations. The percentage of the gross domestic product (the total value of all production) related to agriculture is about 20 percent of the total (Rasmussen, 1989). More than twenty million people work in some phase of agriculture, from researching a gene-spliced tomato, for example, to selling it at the supermarket. However, only 19 percent of the twenty million are producers. The majority of agriculturally related jobs, 52 percent, are in the wholesale/retail trade.

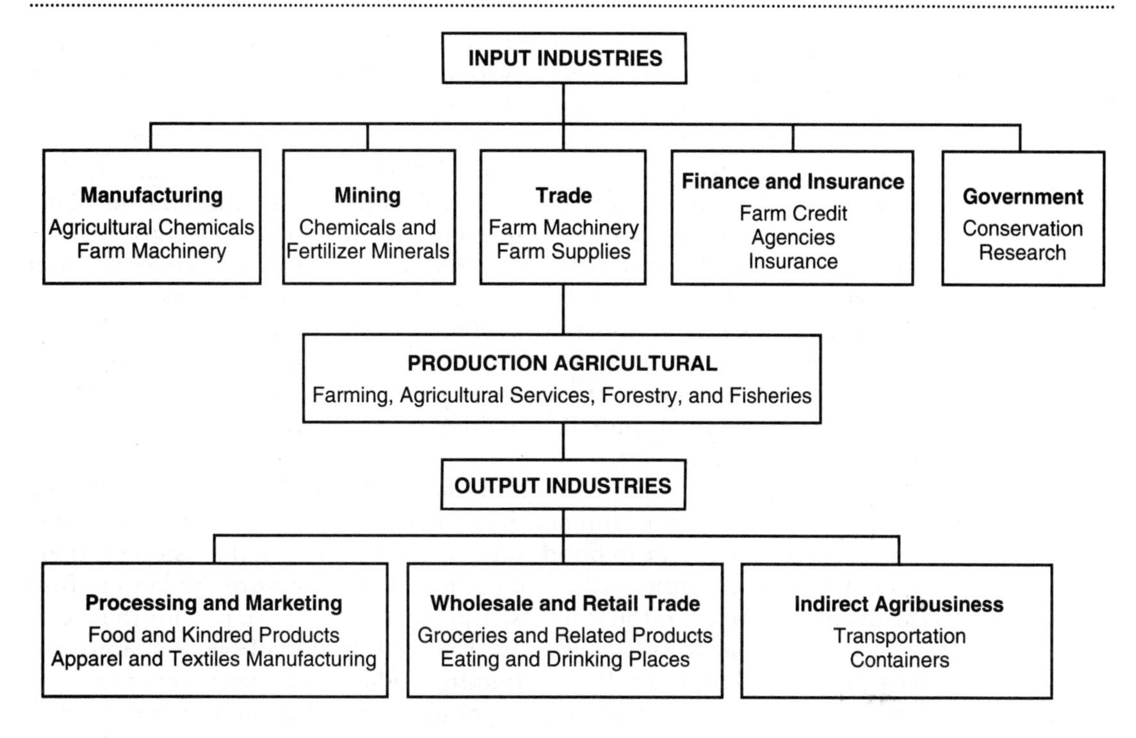

Figure 4–1 The Agriculture Industries. Courtesy USDA.

Education and technology transfer via Extension deserve at least part of the credit for the rise in agricultural productivity and the small percentage (15 to 17 percent) of the American household dollar spent on food (National Research Council 1989, p. 34).

One measure of a nation's development is a decrease in the percentage of its population engaged in basic agriculture. A rise in agricultural productivity contributes to economic development in other sectors. An abundant supply of affordable food and fiber contributes to political stability and also frees workers for employment in other occupations. This is what has happened in the United States. Although the number of acres farmed has remained fairly constant in the last century, fewer and fewer farmers are working the land and feeding an ever-increasing population. Although only 0.5 percent of the farms are classified as very large, they account for 27.8 percent of the sales of agricultural products (National Research Council 1989, p. 34).

Audiences

Primary audiences for the agriculture program area are producers, both large and small. Secondary audiences are those who provide inputs, services, and education for producers. These include bankers; feed, fertilizer, and seed dealers; grain and livestock buyers; grain elevator personnel; agricultural teachers; commodity groups; farm organizations; health officials; and agencies of the USDA.

One USDA agency that works closely with Extension is the Natural Resource Conservation Service (NRCS), formerly the Soil Conservation Service. It helps landowners develop and implement conservation plans and assists with projects such as terraces and windbreaks. Another USDA agency is the Farm Service Agency (FSA), which deals with governmental regulations and funding, see figure 4–2. The FSA includes parts of the former Farmers Home Administration (FmHA) and Agricultural Stabilization and Commodity Service (ASCS). Extension's role in the USDA is education; other agencies provide technical assistance and funding. However, there has been a blurring of these distinctions in recent years as all have taken on some educational roles.

Local personnel often cooperate on projects to benefit clientele. For example, an Extension crop specialist may conduct a tour that prompts participants on the tour to build some terraces to stop soil erosion on their land. A technician from NRCS will come out and measure the land to find the proper location for the terraces, and the FSA will help pay for the cost.

Extension has a legislated commitment to serve all people. However, wise program planning calls for targeting audiences to better serve specific needs. When program planners target agricultural audiences for specific programs, the targeting tends to be by crop or livestock enterprise. Clients are also targeted by age, geographic area, part- or full-time employment, and private or commercial operation. Not all agricultural Extension programs are directed toward the family farmer; for example, horticulture programs are often divided between commercial growers and private home gardeners. Turf grass programs, targeted at commercial growers, are economically important to golf course managers or anyone with responsibility for grassy areas.

The difficulty of reaching a large and wide-spread audience with diverse needs prompted Extension to develop satellite training, telephone answer-lines, radio programs to answer call-in questions, and master volunteer programs. The most successful master volunteer program is the Master Gardener Program.

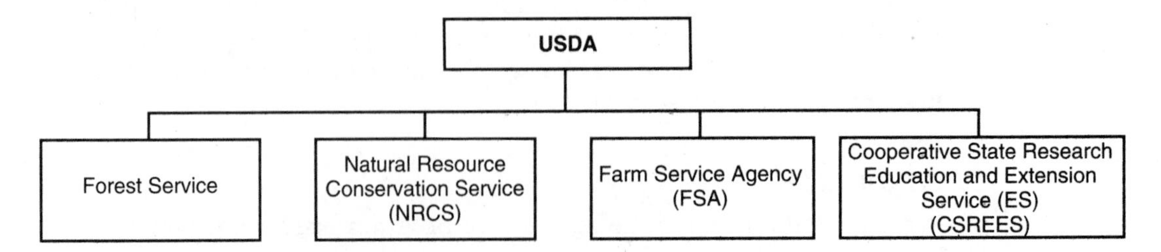

Figure 4–2 United States Department of Agriculture Farm-Related Agencies.

Master gardeners receive approximately forty hours of training and are obligated in turn to spend forty hours educating others. They answer horticultural questions, assist with 4-H gardening activities, sponsor clinics, work with local farmers' markets, and assist with community plantings. The Master Gardener program has been a model for other programs such as Master Sheep Producer and Livestock Master (United States Department of Agriculture 1990, p. 171).

Components and Scope

The scope of agricultural Extension programs is comprehensive; it ranges from A to Z, agronomy to zoology. A new agricultural agent observed:

> *I thought I would be advising farmers about growing corn. Instead, my first office caller was a homeowner with a snake in her basement! Fortunately, we had a bulletin on snakes and mice, and my office assistant helped me locate it (Clete Swackhamer, Appanoose County Agent, Oral Presentation, 1991).*

The university agricultural Extension specialist who wrote the bulletin on snakes and mice was in the animal ecology department. Every discipline related to agriculture is likely to have one or more faculty members who have Extension as a major portion of their responsibility; larger academic departments such as agronomy (crops and soils) and animal science are likely to have several Extension specialists. Agronomy specialists are responsible for providing programs and materials related to crops, such as information on when and how to take soil tests. Entomology specialists provide information on pest management that helps producers decide whether it is wise to spend money to spray insects. For help with marketing crops and livestock, such as use of commodity exchanges, economics specialists are the ones to consult. When a producer is interested in new irrigation equipment or use of alternative energy in dairying facilities, an engineering specialist will have part of the answer with help from crop or animal science specialists, respectively. The above are general categories; a specialist's expertise may be more focused, on sheep nutrition, for example.

A particular region will need people with particular expertise. At Washington State University, an Extension trade specialist plays an important role in agricultural export trade enhancement (Youmans 1994), and Cornell Cooperative Extension in New York recruits staff with dairy science degrees for some of its community educator (county agent) positions.

The local level agricultural educator is the link between the land-grant research and the local people, a person who knows where to get the answers and how to develop programs to meet local needs. A typical day might include a visit to inspect a dying tree, telephone consultations on cotton planting decisions, and recruiting an audience for an up-coming meeting featuring a water quality specialist. An Extension agriculturist is often asked to present programs for existing clubs and groups.

Educational Methods

Agriculturists tend to prefer these educational methods: meetings, tours, radio, magazine articles, newsletters, and individual consultations. As they listen to office callers and make visits to farms, acreages, and lawns, Extension agriculturists develop ideas for reaching others who may have similar problems, such as insect infestations or difficulty complying with federal regulations. Individual contacts generate ideas for group contacts through meetings, radio, newsletters, and newspaper columns.

The use of newer delivery tools such as fiber optics, satellite, and computer programs (Park and Gamon 1995) is increasing. Early uses of computers, beginning in the 1960s, were for farm record summaries, data bases for integrated pest management, and decision-making programs in enterprise management. Later uses have included education on topics such as sustainable agriculture (Sisk 1995).

Crisis situations have prompted the agricultural program area to use a variety of delivery methods. One of the hallmarks of Extension is a quick, knowledgeable response to adverse conditions. The farm crisis of the 1980s spawned innovative Extension efforts to assist farmers and farm families, such as telephone hot-lines staffed by part-time, empathetic people trained to answer calls from stressed rural citizens.

Major Agricultural Programs

Profitability is a major program emphasis for agricultural Extension. Examples of topics related to agricultural profitability are the following:

- Reduction of land in the federal CRP (the Conservation Reserve Program that paid producers to take land out of production)
- Contract production of livestock
- Multiple-generation farm ownership
- Cooperative buying/selling arrangements
- Government farm policy changes
- Financing by nontraditional lenders
- Environmental impact of farming practices

(Iowa State University Extension 1992)

The systems approach, which views agriculture as part of a larger societal system, is essential to profitability. An example of a systems approach, described in a national Extension study (Lippke, Ladewig, and Taylor-Powell 1987), was a range livestock integrated resource management program in Montana.

Natural Resource Programs

Environmental concerns, such as sustainable agriculture and water quality, are related to natural resource programs. In some states natural resources is a part of the agriculture program area; in other states it is a separate program area. Natural resources include water, soil, forests, and wildlife. All of these are affected by agricultural practices such as waste management and type of tillage. At the local level of Extension programming, the divisions seem moot, but at the state level, departments and specialties in specific topics may make it difficult to translate research into user-friendly programs.

Programs that address environmental concerns typically contain input from several program areas. For example, in a pesticide applicator training program taught by an agriculturist, a home economist may talk about how to safely launder the clothing worn during chemical application.

HOME ECONOMICS

The Extension home economics program area is an educational program for families and consumers. Programs fit into three generally recognized categories: nutrition, resource management, and human development. In some states the title for the program area is home economics; in others it is family and consumer science, or family living.

Goals

The goal of the home economics program area is to provide practical, research-based education to strengthen the well-being of individuals and families. Well-being includes physical, social, and emotional factors. The home economics program area focuses on development throughout the life span, management of family finances, and issues in nutrition and health. Helping consumers make informed decisions and enabling families to survive in a changing environment is a part of the goals of the home economics program area.

Audiences

The home economics program area includes all families and consumers in its potential client base. The particular audience targeted is dependent upon the content area and the pressing issues and needs, see figure 4–3. For example, resource management programs sometimes target older families (O'Neil 1993), or nutrition programs target limited resource families (Achterberg, Van Horn, Maretzki, and Matheson 1994).

Although the audience for home economics programs is theoretically very broad, programs are targeted to specific groups. Since the direction of work is toward prevention of problems through education, child-care providers and 4-H/youth audiences are included in the clientele base.

A major thrust is to reach and help families at risk. Following are examples of clients-at-risk:

- Families who live in temporary housing or who spend more than one-third of their incomes for shelter

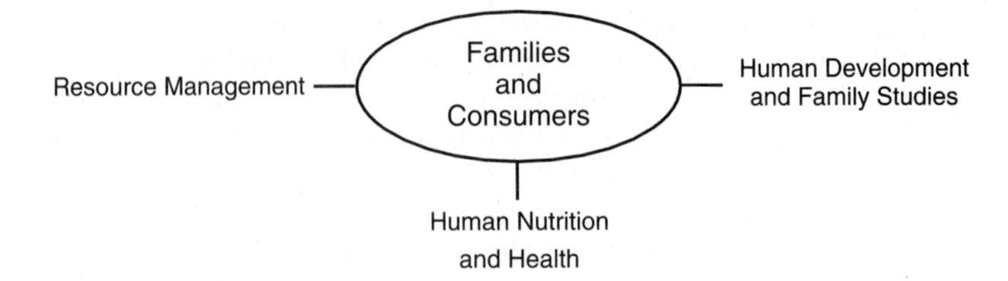

Figure 4–3 Audiences and Content Areas for Home Economics Programs.

- Families who lack food and nutritional knowledge
- Families who can't get medical care and are unable to care for themselves
- Abusive parents and siblings
- Persons with minimal or entry-level job skills
- Families with unpaid bills, a high debt load, no savings, and no credit
- Families whose members are not nurturing
- Families experiencing severe change due to loss
- Teenage parents

Educational Delivery Methods

Extension educators in the home economics program area are expected to work with a variety of community organizations to serve people. They often coordinate their programming with professionals in other agencies and serve as liaisons, facilitators, and role models as they implement programs to serve families and consumers. Their interagency work is extensive, as they work with a wide variety of health and human service groups, such as councils on aging, child-care networks, and community mental health groups.

Home economists use volunteers and advisory councils to the fullest. The first home agents gave demonstrations to both women and girls and almost from the beginning formed clubs to reach more of their clientele. Family community education clubs (formerly homemaker clubs) send volunteers to receive training from Extension staff; they then return to their clubs and present a lesson. Most states have homemaker clubs, and they may use creative methods to reach members. For example, "Mailbox Homemakers" in Illinois receive information in the mail rather than at meetings. Another method to reach clubs consists of packaged programs, a set of materials that a volunteer may pick up at the Extension office and share with others in the group.

Figure 4–4 illustrates the various groups and methods home economists use to reach their audiences. With limited resources for personnel, educa-

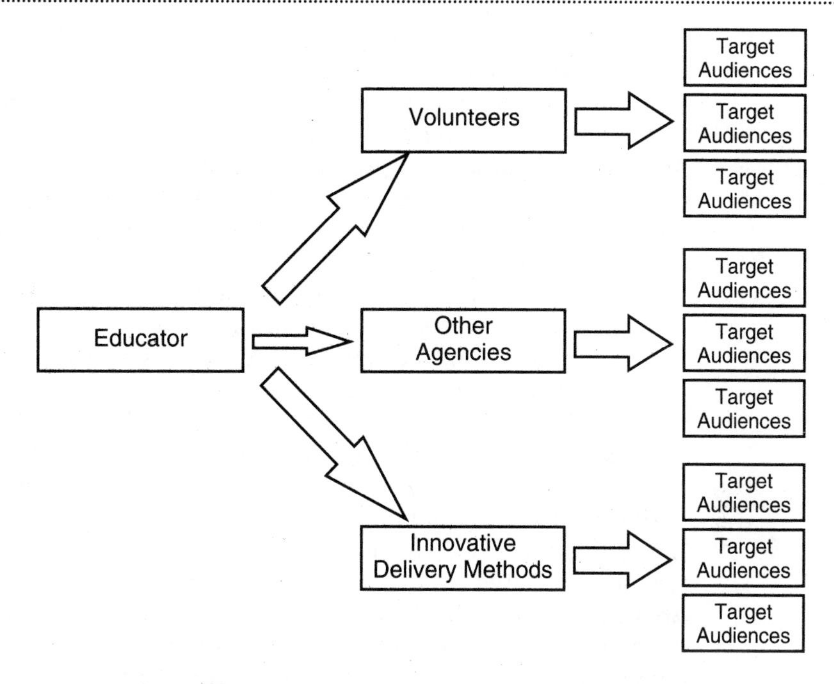

Figure 4–4 The Multiplier Effect

tors in the home economics program area are constantly challenged to find innovative delivery methods. They were among the first to deliver programs via television and satellite. The leadership, counsel, and educational materials home economists provide to professionals and volunteers help to multiply their impact many times over.

Program Scope

As society has changed, the scope of the home economics program area has broadened and priorities for programming have adjusted. Programs fit into three general areas: human development, family resource management, and human nutrition. The human development area has a special focus on children from birth to five years of age and the elderly. Family resource management includes planning for food, clothing, housing, and consumer education. Nutrition includes health, wellness, and food safety.

Families of juvenile offenders are an example of a target audience that home economics Extension serves. In Ohio, Extension educators developed a six-week parenting course and reduced repeat offenses (Jackson 1993). Parenting education reaches child-care providers as well as parents and guardians. A satellite program on quality child care reached a nationwide audience with information for those who work with young children. (Boehn, Hendricks, and Steffens 1993).

The mission to improve the well-being of families sometimes takes a different focus than working directly with family members. For example, in Missouri, the textiles and clothing Extension program redirected its efforts to manufacturers and retailers to help apparel manufacturers improve their competitive positions. Extension made the effort because the apparel sector had production sites in 82 of Missouri's 114 counties, and farm families were increasingly dependent on nonfarm earnings (Dickerson, Dillard, and Froke 1994).

Another innovative delivery method was used in Ohio to teach stress management and discipline techniques to parents of young children. Workers at fast food outlets handed out a set of cards with each kiddie meal. Each of the cards had one suggestion for an activity. In an evaluation of the program, home economists found that 68 percent of the parents had read at least one of the cards (Syracuse, Kightling, and Conone 1993).

The Expanded Food and Nutrition Program (EFNEP), a national nutrition education program for limited-resource adults and youth, began in the 1960s with a Congressional mandate and "soft money" funding (money allocated on a year-by-year basis). EFNEP helps limited-resource urban and rural families acquire knowledge and skills to improve their diets. A unique feature of the program is the hiring of paraprofessionals, people without college degrees, who are from the targeted communities. EFNEP program aides visit homes and demonstrate selection, preparation, and serving of nutritious family foods. Extension professionals develop training materials, bulletins, and leaflets targeted at the clientele and supervise and train EFNEP program aides.

4-H/YOUTH PROGRAM

The 4-H/Youth program is a voluntary educational program that supplements formal school education. It provides real-life experiences and an opportunity for youth to plan their own learning, with parents and other adults to guide them and evaluate their accomplishments.

Professionals, paraprofessionals, and volunteer adults help reach both urban and rural youth. Educational delivery methods include clubs, short-term special interest programs, camps, and programs before, during, and after school.

The 4-H/Youth program is the most widely recognized of the Extension program areas (Warner and Christenson 1984). More than eighty countries around the world have similar programs and similar words for the four Hs of the 4-H clover: head, heart, hands, and health. All programs are helping young people gain skills needed to become responsible, productive members of society.

Mission and Goals

In 1991, a National 4-H Strategic Planning Conference defined the mission of 4-H as helping youth "become productive citizens," or in other words, "self-directing, contributing members of society" (Iowa State University 1992). The national mission calls for a "supportive environment for culturally diverse youth and adults to reach their fullest potential."

The 4-H/Youth program has an experiential philosophical base. Young people learn by doing and by becoming self-directing. This means they set their own goals, make plans to meet them, and keep records to document their achievements.

Skills that youth learn include content-based skills and living skills. The objectives of building character and citizenship are embodied in the following living skills:

- decision making
- leadership
- citizenship
- communication
- self-esteem
- coping with change
- learning how to learn

(Iowa State University Extension 1989)

Content skills are subject-matter skills that rely on the resources of the land-grant universities. Examples of content-based skills are choosing a toy to use in baby-sitting or using a computer program to keep track of lawn-mowing business expenses.

Symbol and Pledge

The USDA adopted the four-leaf clover, a symbol of good luck, as the official 4-H club symbol in 1911. In the 1920s the name, 4-H, and the 4-H pledge came into being. The clover is green, symbolizing the color of growth and the Hs on each clover are white, symbolizing the purity of youth. The clover symbol cannot be used commercially without permission from Extension administration at the national level. The pledge is recited with appropriate gestures at club meetings and 4-H activities, see figure 4-5.

Organizational Structure

The organizational structure of the Extension youth education program includes clubs, short-term activities, school enrichment, and other delivery methods. Since 1980, the number of young people in organized clubs has decreased while the number participating in other delivery methods has increased, see figure 4-6. Organized community 4-H clubs are ongoing groups of youth who meet regularly with adult leaders. Short-term activities include camps, conferences, and special interest groups who meet for a short time. School enrichment programs offer hands-on activities taught by the teacher or visiting Extension personnel.

I pledge
My head to clearer thinking,
My heart to greater loyalty,
My hands to larger service,
My health to better living,
For my club, my community,
my country and my world.

Figure 4–5 4-H Clover and Pledge

Community clubs and project clubs are an important delivery mode for the Extension youth education program. The community club is organized around a township or school district, and the project club is organized around a common project, such as photography. The clubs, typically serving youth nine to nineteen years old, often meet in community buildings,

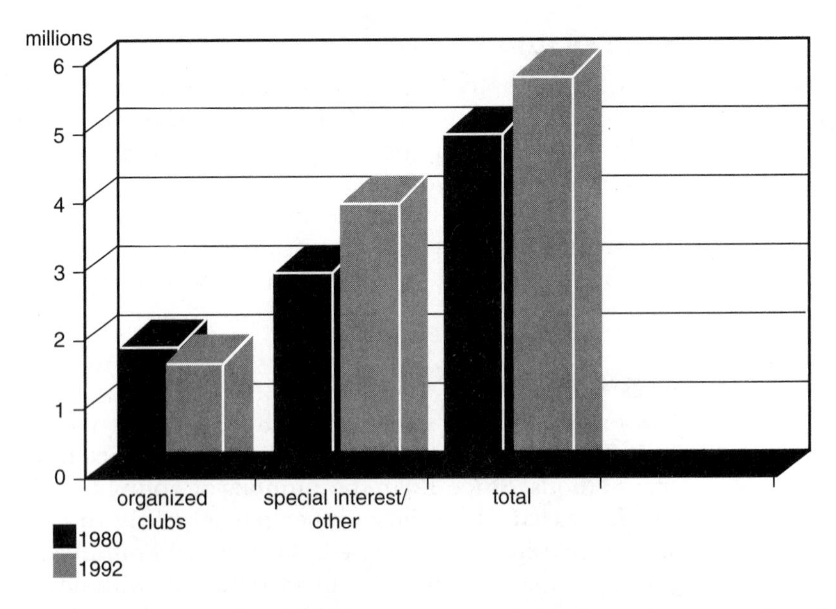

Figure 4–6 National 4-H Enrollments 1980 and 1992, by Type of Educational Delivery Mode. 4-H Youth Development Enrollment Report, ES-USDA, 1992, 1980.

schools, or homes once a month, and have a business meeting as well as an educational program. The educational program usually consists of talks and presentations by club members and may also include adult speakers. Some states have a tradition of small clubs with as few as six members and one or two adults serving as volunteer leaders; other states have large clubs that may have fifty or more members. The clubs may meet in the evening, after school, or during the school day. Parents, teachers, other interested adults, or teens serve as leaders.

Community 4-H clubs provide their members with an opportunity to feel a part of a group and to learn social skills, parliamentary procedure, and leadership skills. Many clubs perform community service activities, such as planting flower beds, taking pets to visit the elderly, and helping disabled children ride horseback.

Boys and girls who join clubs that meet regularly agree to work on projects at home under adult supervision, to engage in activities with volunteer leaders, to set goals and keep records, and to explain and show to others what they have done.

The most visible demonstration of their accomplishments is at fairs and shows, where they exhibit their projects. Fair exhibits are grouped into classes such as visual arts, horticulture, livestock, nutrition, clothing, and many others. Judges evaluate the projects and award blue, red, and white ribbons, with blue being the best. Top-ranked projects often receive trophies and/or purple ribbons.

Other delivery modes in addition to community clubs include urban programs, community development programs for youth, Expanded Food and Nutrition Programs for children with limited resources, residential camps, day camps, tours, educational television, short-term special interest groups, and school enrichment programs. All 4-H/Youth programs have subject-matter content and goals related to living skills.

Target Audiences

In many states, more youth are involved in other educational delivery modes, such as school enrichment, than in 4-H community clubs. In the school enrichment program, an Extension Expanded Food and Nutrition Program aide employed specifically to work with nutrition programs in urban schools might take ingredients to the classroom and help the students make a nutritious snack. The lesson would include an explanation of the importance of eating fruits and vegetables.

The growing number of youth involved in nontraditional programs is partly due to the increasing urbanization of today's 4-H/Youth programs. Nationwide, a small proportion of the total 4-H/Youth audience resides on farms, see figure 4–7. Over one-half live in communities of 10,000 or more.

Youth at Risk

Extension's youth-at-risk initiatives focus on prevention and intervention rather than treatment. Youth at risk are vulnerable because of poverty,

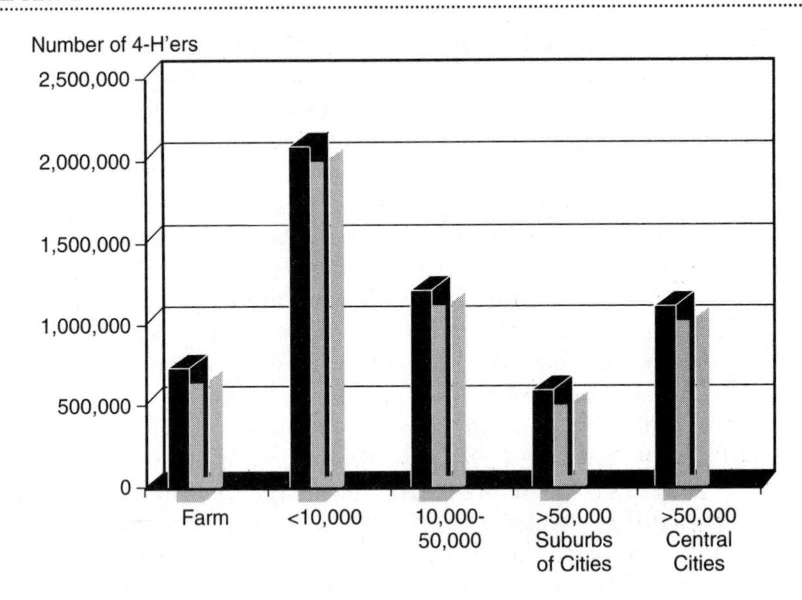

Figure 4–7 Number of 4-H'ers Enrolled by Place of Residence 1994

lack of parental and community support, and negative peer pressure. Support from private citizens, grants, and cooperation with other agencies are integral components of youth-at-risk programs. The following are examples of programs:

- Illinois: Computer-Assisted Learning was a reading and science literacy program in the Alton/East St. Louis area for high school-age youth who were truants and dropouts.
- Minnesota: Hmong, Hispanic, and Laotian families in southeastern Minnesota were helped to adapt to a new culture though Project FINE (Focus on Integrating Newcomers into Education). Youth in kindergarten through eighth grade participated with their families in enrichment and cultural awareness activities.

Programs for Younger Children

Early efforts at programming for younger audiences (kindergarten through third grade), used a simplified version of 4-H with names such as Clover-buds, mini-4-H, and Sprouts. In the 1970s delivery methods were day camps, weekly meetings, and family 4-H.

Extension educators realized that a new program was needed to suit the developmental characteristics of younger children; therefore, a national Extension conference in 1995 focused on kindergarten through third grade (K-3). Participants developed policies and guidelines for age-appropriate materials to ensure that K-3 programs were not just a simplified version of 4-H, but geared specifically to that developmental stage.

Instead of the competition that is an important part of 4-H for older members, the younger children have cooperative activities. The whole program is designed to reach families and involve communities in getting children off to a good start.

4-H Projects

A 4-H project is a self-directed project in a content-based area. Researchers have found that projects help 4-H'ers learn life skills while they are learning content skills (Gamon and Dehegedus-Hetzel 1994).

4-H'ers get help with their projects from project leaders, family members, and project manuals. Project manuals include age-specific study material and suggested activities to enhance learning about the subject. 4-H project manuals are provided at low cost by national donors to the 4-H program or developed on a state or regional basis. In addition to member manuals, there are leaders' guides for adults who are volunteering with the project.

A 4-H'er may exhibit part of his or her project at a local show or county fair and receive ribbons or awards. Some 4-H projects, such as livestock and "learn and earn" projects, are profit-making ventures for youth.

Some of the first 4-H projects involved raising corn, pigs, and tomatoes. Today there are over a hundred projects and areas such as computers, citizenship, and health have been added.

4-H Awards and Recognition

Contests and competitions have been a part of 4-H almost since its inception and the early corn contests. Recently, more emphasis is being placed on cooperative efforts. A national recognition model emphasizes that recognition is desirable for five different kinds of achievement: participation, progress toward self-set goals, achieving standards of excellence, excelling in peer competition, and working cooperatively (Parsons 1995), see figure 4–8. The rules, the process a 4-H'er must go through to achieve the recognition, and the type of award differ for each of the five categories. For example, a 4-H'er will be one of many who participate at the local level at a young age, but one of a few who achieve national recognition at the end of a 4-H career. Examples of recognition for participation are green ribbons, t-shirts, and certificates. Rewards for individual progress toward self-set goals may be small gifts of 4-H memorabilia and/or positive comments.

4-H Activities Beyond the Local Level

Recognition, especially for older 4-H'ers, may consist of trips, awards, or honors beyond the local level. At the county or parish level, 4-H'ers may be a part of councils, committees, or officer groups. State conferences, camps, exchanges between states, and citizenship programs at the National 4-H Center in Chevy Chase, Maryland, have fostered friendships and in-

Figure 4–8 ***The National 4-H Recognition Model. Jerry Parsons 1995. Recognitions in Youth Programs. National 4-H Council, Chevy Chase, MD.***

creased leadership skills of the young people involved. The 4-H Congress and the National 4-H Conference provide leadership opportunities for 4-H'ers as well as opportunities for donors to recognize outstanding 4-H'ers for their efforts. International experiences are available for older 4-H'ers, leaders, and Extension educators. The National 4-H Council sponsors the International Four-H Youth Exchange (IFYE) and other exchanges between American 4-H members and leaders of similar programs in other nations.

The 4-H Professional Research and Knowledge Base

The 4-H Professional Research and Knowledge base (PRK) was established in 1986 and revised in 1994. Its purpose is to define the knowledge and research base of 4-H youth development programs.

4-H professionals are responsible for managing 4-H/Youth programs and have a number of important roles, including the following:

- supporting local clubs
- technical advising
- working with volunteers
- youth counseling
- planning, implementing, and evaluating educational programs
- public relations and recruiting audiences
- fund raising with private donors

To help professionals with their roles and to provide a framework for research efforts, the panel of experts who worked on the PRK identified five key knowledge areas: communication, educational design, youth development, program management, and volunteerism. Also, the national 4-H staff created a repository of materials related to 4-H youth development at the National Agricultural Library in Beltsville, Maryland, and "4h prk" is included on the CD-Rom edition of the AGRICOLA data base available at university libraries. States use the five key knowledge areas as a guide for in-service and preservice programs, as a model for hiring and performance standards, and as expertise areas for 4-H specialists.

COMMUNITY DEVELOPMENT

The community development program area is different from the other three major program areas in the following aspects:

- It was not mentioned in the original Smith-Lever Act of 1914.
- It is not as likely to have a full-time Extension educator at the county level with community development as his or her main emphasis.
- It is much smaller in terms of staffing and funding.
- It is process- rather than content-oriented.

These features might seem to make it less important; however, in 1993, the USDA declared community development to be a priority for programming emphasis.

Goals

The goal of the community development program area is to improve the social and economic well-being of communities though group action. A paraphrase of the mission is "help communities help themselves." Community development programs are defined as much by their audience as by their content. While the other program areas deal with individuals or families or businesses, the community development program works mainly with groups.

Audiences

Community development programs were mandated by the federal Rural Development Act of 1972 to serve rural communities, defined by the Census Bureau as all incorporated areas with fewer than fifty thousand people. If the population is at least fifty thousand, an area is considered urban. Within rural communities, city or town officials such as city clerks, community organizations such as Chambers of Commerce and community betterment groups, and key leaders are target audiences for community development programs.

Components of the Community Development Program

The four main components of Extension community development programs are leadership, public policy, economic development, and commuity services and facilities. Census figures and social and economic surveys are important tools for all four components.

Leadership programs help people become more effective in their community settings. Extension educators train leaders in teamwork and group dynamics and help them find resources. Participants learn how to use their new skills to solve community problems.

An example of a leadership program is MI/LEAD, a program jointly established by Minnesota and Iowa that provides in-depth training for small groups of emerging leaders and encourages them as they develop projects to help their communities.

Public policy education teaches those who will be affected by a policy how to deal with controversy. "If an issue exists or is emerging, controversy over the objective, alternatives, and interpretation of the consequences is to be expected" (House 1988). Examples of topics addressed by public policy education are the effects of bond issues and tax policies on a community. Whereas most of the community development programs deal with facilitating change, the public policy aspect alerts citizen groups to the effects of change. Public policy programs typically emphasize scientifically developed information and use logic and data in analyzing alternatives.

Economic development programs are centered around jobs, businesses, and industry. Programs for managers and owners of main street businesses fall into this category. Specialists offer specific training in sales strategies, customer satisfaction, and business retention. Also, Extension specialists in this area study the effects of new businesses on a community and suggest ways businesses could position themselves to be competitive.

The **community services** and facilities aspect of the community development program area includes programs for local government, the arts, health care, housing, and recreational opportunities. In these, the role of the Extension educator is to train councils and personnel and to help communities solve their problems. Content may include financial management, personnel management, and decision-making techniques. Programs in this area are process oriented in that the educator facilitates the direction of group energy toward some common goal, such as a new swimming pool, or housing for the elderly.

Community Development Processes

Extension educators try to follow a systematic process in assisting groups to reach their goals. The social action process developed by Beal (1964) is a classic process for solving community problems. The Cluster Action Plan (Foley and Tait 1993) is a simplified version of the social action process, see figure 4–9. It is accompanied by a series of questions to help guide citizens through a community involvement project.

Extension specialists in community development tend to have backgrounds in sociology, economics, political science, and community and regional planning. When Lackey and Pratuckchai (1991) surveyed community development specialists and asked them to rate competencies

1. **Define Problem**	2. **Analyze Situation**	3. **Seek Approval**
Define Problem/Opportunity	Priority	Decision Makers
Cause or Symptom	Gather Information	Contacts
Desired Outcome	Helping Forces	Opponents
Who Benefits	Hindering Forces	Strategy
Initiators	Alternative Strategies	
4. **Develop Strategy**	5. **Support**	6. **Evaluate**
Select Strategy	Awareness	Successes
Action Team	Motivation	Failures
Action Steps	Recognition	New Resources
Resources		Spin-Offs
When		
Who		

Figure 4–9 Cluster Action Plan. Mary Foley and John Tait 1993. Iowa State University Extension, Ames, IA.

needed in the profession, skills in organization, analysis, leadership, and human relations were rated most important.

Extension sociologists Beal and Bohlen developed a model for the adoption of new ideas, a model that has been widely researched (Rogers 1995). The adoption-diffusion model was originally developed to explain the educational processes that led farmers to accept new ideas, but it has been widely used in other areas as well. The adoption-diffusion model is discussed in Chapter 6.

INTERDISCIPLINARY PROGRAMS AND OTHER PROGRAM AREAS

Although cooperation among program areas has always existed, interdisciplinary programs became more important beginning in the 1980s. The concept of issue programming encouraged interdisciplinary programming efforts. In issue programming, input from clientele at all levels—county, state and national—helps to identify the issues for Extension to address.

The issues form a focus for local programs to meet local needs. For example, water quality is not only a community and agricultural concern, but also one that youth and family-living programming can address. How local programs develop depends upon the wishes and resources of the people in the local community.

Other Program Areas

Energy programming was at one time a separate entity, designated as Energy Extension, but the duplication vanished after the energy crisis of the early eighties faded. Now all program areas participate in programs related to energy.

Another program with the potential for duplication of efforts was the Sea Grant program, which was developed in 1966 to promote marine programs and programs around the Great Lakes. Most of the states integrated it into Cooperative Extension.

Some states have "University Extension," which includes program areas such as business, industry, health, and science, in addition to Cooperative Extension. The trend toward issue programming encouraged a trend toward the use of university resources beyond the traditional agriculture and home economics base.

Offering credit and noncredit courses to people off-campus is an outreach effort for colleges and universities that is called Continuing Education. States structure the features and organization of Continuing Education in various ways, and courses offered vary widely in their content, cost, length, and structure. Computer training, nutrition training for school cooks, and sessions for bankers are examples of courses that might be offered.

SUMMARY

Each one of the program areas has different features. Agriculture has the largest budget for personnel. It addresses both environmental concerns and profitability. Audiences include both producers and those who provide input, services, and education for producers. Although the number of people who list farming as an occupation has dropped below 2 percent, about 18 percent of the labor force works in an occupation related to agriculture.

The home economics program area has the most general audience—families and consumers. A major thrust of the program is families at risk economically or socially. Extension educators work through other agencies and volunteers to multiply their efforts.

The 4-H/youth program area is the most well-known of the program areas (Warner and Christenson 1984). It reaches a school-age audience and uses a wide variety of subject-matter content. New delivery methods teach content skills and living skills to rural and urban youth.

Community development is the newest, the smallest, and the fastest-growing program area. Its mission is to help communities improve their social and economic well-being. The four main components of the community development program area are leadership, public policy, economic development, and community services and facilities.

An emerging trend in programming is interdisciplinary programs, which use parts of two or more of the four main program areas (agriculture, home economics, 4-H/youth, and community development). This is consistent with the emphasis on issue programming, which means developing programs based on important emerging needs of people. The expertise to meet those needs might come from one or more of the four main program areas or from other sources.

DISCUSSION QUESTIONS

1. How is the agriculture program area changing?
2. What are the negative and positive implications of targeting an at-risk population in the home economics program area?
3. Considering the trends, what might the mission, audience, and delivery methods of 4-H look like in the future?
4. Assume that an aging population is a problem for a small town. How might the four components of the community development program area address this problem?
5. What might be the constraints of an interdisciplinary approach? At the local level? State level? Federal level?

REFERENCES

Achterberg, C. L.; Van Horn, B., Maretzki, A., and Matheson, D. 1994. Evaluation of dietary guideline bulletins revised for a low literate audience. *Journal of Extension* (on-line) 32. Available Internet: Almanac@joe.org Message: send joe december 1994 research.

Beal, G. M. 1964. Social action: Instigated social change in large social systems. In *Our changing rural society,* J. H. Copp, ed.. Ames, IA: Iowa State University Press.

Boehn, L., Hendricks, P. A., and Steffens, P. E. Summer 1993. Training for quality child care. *Journal of Extension* 31(2):13–14.

Cooperative State Research Service. July 1993. *Dynamics of the research investment: Issues and trends in the agricultural research system.* Washington, D.C.: United States Department of Agriculture, Cooperative State Research Service.

Dickerson, K. G., Dillar, B., and Froke, B. Spring 1994. Assisting rural economies and families. *Journal of Home Economics* 86(1):55–60.

Extension Committee on Organization and Policy. 1990. *Strategic directions of the Cooperative Extension System.* Washington, D.C.: Extension Service, USDA, Extension Committee on Organization and Policy, National Association for State Universities & Land-Grant Colleges.

Economic Research Service. 1987. *U.S. food and fiber system: Agriculture economic report.* NFR-37 and No. 566. Washington, D.C.: USDA, Economic Research Service.

Foley, M., and Tait, J. 1993. *Cluster action plan.* TLT 12. Iowa State University, Ames, IA: Extension Publications Distribution.

Gamon, J. A., and Dehegedus-Hetzel, O. P. June 1994. Swine project skill development. *Journal of Extension* [on-line] 32. Available Internet: almanac@joe.org Message: send joe june 1994 research 5.

House, V. W. 1988. Education for public decisions. In *Working with our publics* (Module 6), E. J. Boone, ed. Raleigh, NC: North Carolina State University, Raleigh.

Iowa 4-H Statistical Report. June 1993. Ames, IA: Youth and 4-H Office.

Iowa State University Extension. 1989. *Iowa 4-H teaches life skills.* (4-H4A). Ames, IA: Extension Publication Distribution.

Iowa State University Extension. 1992. *Trends in Iowa agriculture: Opportunities for extension programming.* Ames, IA: Iowa State University Extension

Iowa State University. December 1992. *Iowa Youth and 4-H: Vision, mission, values.* 4-H4. Iowa State University, Ames, IA: Extension Publications Distribution.

Iverson, R. M.; and Rohs, F. R. 1994. A team approach to agricultural and extension education in Georgia. *The Agricultural Education Magazine* 66(11):12–14.

Jackson, B. J. Fall 1993. Juvenile court parenting program. *Journal of Extension* 31(3):35–36.

Lackey, A.S., and Pratuckchai, W. 1991. Knowledge and skills required by community development professionals. *Journal of the Community Development Society* 22(1).

Lippke, L A., Ladewig, H. W., and Taylor-Powell, E. 1987. *National assessment of Extension efforts to increase farm profitability through integrated programs.* College Station, TX: Texas Agricultural Extension Service.

National Research Council. 1989. *Alternative agriculture.* Washington, D.C.. National Academy Press.

O'Neill, B. B. Summer 1993. Gaining "repeat" customers for Extension. *Journal of Extension* 31(2):33–34.

Park, S. Y.; and Gamon, J.A. 1995. Computer use, experience, knowledge, and attitudes of Extension personnel: Implications for designing educational programs. *Proceedings, National Agricultural Education Research Meeting,* Denver, CO.

Parsons, J. 1995. *Recognition in youth programs: Helping young people grow.* Chevy Chase, MD: National 4-H Council.

Rasmussen, W. D. 1989. *Taking the university to the people: Seventy-five years of Cooperative Extension.* Ames, IA: Iowa State University Press.

Rogers, E. M. 1995. *Diffusion of innovations.* 4th ed. New York, NY: The Free Press.

Sisk, J. G. 1995. *Extension agricultural agents' perceptions of sustainable agriculture in the southern region of the United States.* Unpublished doctoral dissertation. Baton Rouge, LA: Louisiana State University.

Syracuse, C. J., Kightling, D.Y., and Conone, R. Fall 1993. Teaching parenting at McDonald's. *Journal of Extension* 31(3):36–37.

True, A.C. 1929. *A history of agricultural extension work in the United States, 1785–1925.* USDA Miscellaneous Publication No. 15. Washington, D.C.: United States Department of Agriculture.

United States Department of Agriculture. 1990. Americans in agriculture: Portraits of diversity. *Yearbook of Agriculture.* Washington, D.C.: USDA.

Warner, P. D. and Christenson, J. A. 1984. *The Cooperative Extension Service: A national assessment.* Boulder, CO: Westview Press.

Youmans, D. 1994. *Impacts of an extension foreign trade program built on mentoring and teamwork.* Paper presented at the tenth annual conference of the Association for International Agriculture and Extension Education. Arlington, VA.

CHAPTER

"We must ask where we are and whither we are tending."
—Abraham Lincoln

Developing Extension Programs

KEY CONCEPTS

- Major components of program development
- Steps in program planning
- Role of Extension personnel in program planning
- Ways to involve people in the program-planning process
- Current issues in program planning

WHAT IS A PROGRAM?

Extension education is an intentional effort to fulfill predetermined and important needs of people and communities. Single events or activities do not result in the types of behavioral change necessary to accomplish this mission. The word *program* refers to the product resulting from all activities in which a professional educator and learner are involved (Boyle 1981). An example may help clarify this definition. A single activity such as a field day of corn performance trials may not be sufficient alone to help a farmer change tillage practices. But a series of activities such as field days, minimum tillage seminars, dialogue among farmers about using the practices, resources such as computer programs that compare operational costs, and publications about equipment for minimum tillage may be linked together to reach the intended results. The activities or educational methods are all linked by the educational objective to teach farmers about tillage options and factors involved in selecting the most appropriate option for the farm operations. One of the most exciting and fulfilling responsibilities in Extension work is the implementation of a successful program.

Program planning is a process designed to bring about effective programming. This process may be viewed as a system of interrelated parts, all of which work together to achieve defined goals (Conklin and Spiegel 1992). The program-planning process includes identification of needs, planning, instruction, program promotion, evaluation, and reporting (Boyle 1981). Boone (1985) describes program planning as a comprehensive, systematic, and proactive process that results in actions to facilitate changes in behavior of learners and the environment in which they live. The terms *program planning* and *program development* are frequently used interchangeably. The Extension Committee on Policy (1974) defined Extension program development as:

> *a continuous series of complex, interrelated processes which result in the accomplishments of the educational mission and objectives of the organization.*

A review of models from adult and Extension education identified three main components of the program-development process: planning, design and implementation, and evaluation and accountability (Sork and Buskey 1986). These components or processes are linked together in specific ways to result in successful programming, see figure 5–1.

Planning includes several steps that influence the formation of a program: identification of goals, determining needs, setting program priorities, identification of target audiences, and development of program objectives. Planning creates a map for the program that addresses who, what, when, where, why, and with what resources.

Design and implementation builds on the planning to include selection and development of program content, selection and/or development of program delivery methods and resource materials, and creation of time lines for program implementation and evaluation, including putting a program into operation.

Evaluation includes the planning and implementation of procedures to measure various dimensions of program success and impact. Monitoring and evaluating the program provides information for further action. As a result of data, observations, and other sources of feedback, future decisions will be made. How to plan and conduct program evaluation is discussed in depth in Chapter 8.

KEY PEOPLE IN THE PROGRAM-PLANNING PROCESS

In the Life Cycle Model of Program Development, Mayeske (1994) identifies two key groups of people involved in the program-planning process: the guidance team and stakeholders. These groups of people can be described in several ways. **Stakeholders** are defined as people who have a

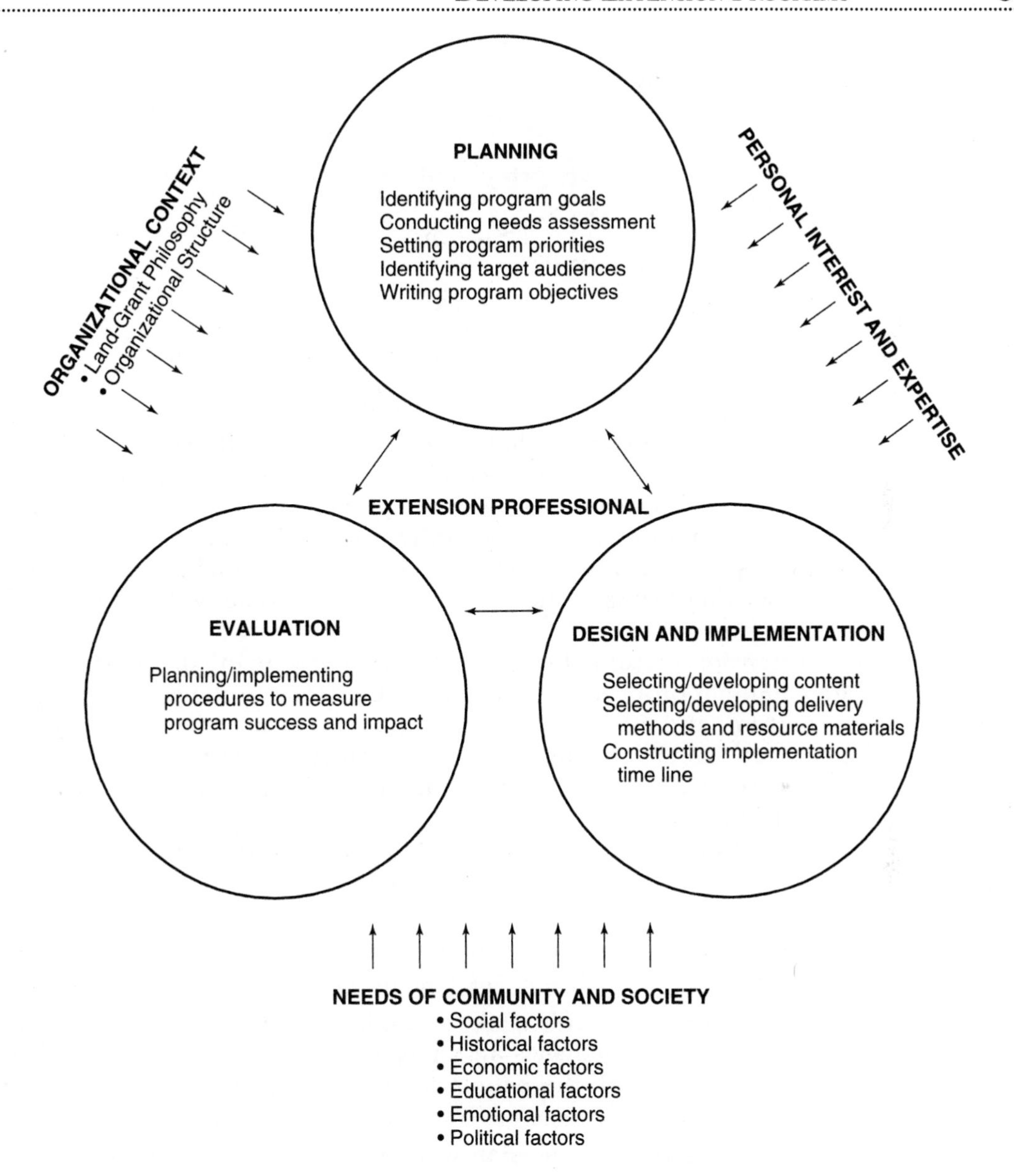

Figure 5-1 ***A Basic Program Development Model.***

vested interest in the program. That can include many individuals: people who participate in the program, Extension personnel, Extension organizational leaders, community leaders, individuals and/or organizations providing funding or other resources to the program, other community professionals who co-sponsor a program, and potential participants in a program, legislators, etc., see figure 5-2. The list may be different depending upon the nature of the program.

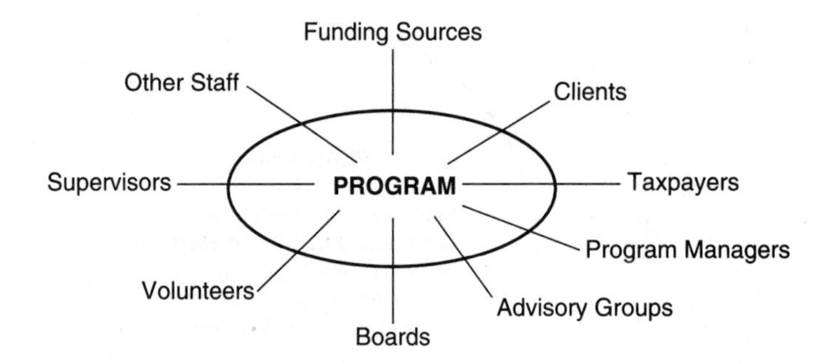

Figure 5-2 Stakeholders in Educational Programs.

The **guidance team** includes the people who will manage the various stages of the planning process. This team is not limited to Extension personnel. Often various stages of the process involve many of the stakeholders, including volunteer coordinators and teachers or community coalitions interested in particular problems. Among Extension personnel, programming teams include many types of people with different roles and responsibilities.

Extension personnel play a key role in organizing the most appropriate combination of people to provide guidance to the planning, implementation, and evaluation of programs. From a review of many adult education models, Boyle (1981) has identified four major roles for the programmer in program development: analyst, stimulator, facilitator, and encourager.

Analyst: The Extension worker in this role is critical for reviewing a situation and helping people diagnose the problem or need.

Stimulator: The Extension worker serves as a stimulus to motivate people and keep the program-planning process going.

Facilitator: This role is like a link in a chain. The Extension worker establishes an environment for learning and helps to link the people to the appropriate knowledge or resources to address the problem or need.

Encourager: The Extension worker helps people to feel comfortable in the learning situation and promotes development of the individual's full potential.

FACTORS INFLUENCING PROGRAM DEVELOPMENT

Many factors influence the planning process and the final product. It is not simple, but they all must be considered and integrated for the plan to be useful. Several important factors to consider include the organizational context, the needs of the community and society, political influences, and personal interest and expertise.

The Extension Organizational Context

Extension philosophy stresses an educational mission that emphasizes practical knowledge, linkage to research, the use of the hands-on approach, and programming in a nonformal or nonschool setting.

Extension's organizational structure also impacts program planning in many ways: funding sources may require specific program goals or methods, staffing impacts programming expertise and responsibilities, geographical areas of responsibility impact who program participants may be, and policies may govern some types of programs. For example, an organizational mission and vision that emphasizes teamwork may affect how staff work together throughout the program-planning process.

The Cooperative Extension System, like most other organizations, establishes long-range plans for programming. These plans guide the distribution of resources such as funds for programming, specialized program staff, publications and audiovisual resources, special grants, etc. This in turn establishes an environment or background for local programming.

Extension establishes program directions at both the federal and state levels. To facilitate long-range planning, the federal level Cooperative State Research, Education, and Extension System (CSREES) requires each state to develop a Four-Year Plan of Work. Annually, the state reports on programming accomplishments and revises its Plan of Work. This plan is based upon program priorities determined from Extension needs assessment throughout the state.

CSREES has outlined programming directions as base programs and national initiatives. **Base programs** are major educational efforts central to the mission of Extension and common to most units. Base programs are ongoing and involve many discipline-based and multidisciplinary programs. They serve as the foundation of Extension programming. Current base programs include Agriculture, Community Resources and Economic Development, Family Development and Resource Management, 4-H and Youth Development, Natural Resources and Environmental Management, and Nutrition, Diet, and Health. **National initiatives** address the most current, significant and complex issues that Extension has the potential to impact. National initiatives arise from the base programs and received special emphasis for a relatively short period of time. Some examples include Food Safety and Quality, Water Quality, Plight of Young Children, and Communities in Economic Transition.

Each state identifies program priorities on a regular basis. Frequently major needs assessments and program prioritization efforts coincide with the four-year program-planning cycle. Many states use a base program concept to establish program priorities in the areas of agriculture and natural resources, 4-H/youth development, home economics, and community development. Others follow the federal example of including initiatives or interdisciplinary program priorities that are current high-priority concerns, such as community leadership development, waste management, or food safety.

"SHEEEP" Factors

"SHEEEP" is an acronym for the social, historical, economic, educational, emotional, and political factors that may impact program planning (Mustian, Liles, and Pettit 1988). These factors may be apparent from a societal perspective, an organizational perspective, or a local community perspective. The successful programmer needs to understand each of these and their impact on program planning.

Social factors include demographic characteristics and trends in society or a community and cultural characteristics that influence how people work together. The following questions can help identify social factors:

- What are the demographic trends globally, nationally, and locally?
- Are global and national trends evident locally?
- What demographic characteristics impact the program being planned? (age, race, ethnic background, sexual orientation, disabilities, gender, religion, military experience, family structure, rural/urban setting, etc.)
- What major social issues are apparent in the area?
- What are the patterns of participation by people who live and work in the community—where do they travel to shop, attend activities, etc.?

Historical factors may include programs that were previously conducted by Extension educators, the results of those programs (both positive and negative), the way Extension professionals interacted with the public in the past, and major events that have occurred or continue to occur in the community.

The **economic factors** of the geographical area or the community can impact program planning. The following questions can help identify economic factors:

- Is the community supported by an agricultural base, industrial base, or a combination?
- What percentage of the population is affiliated with each sector?
- What is the average household income or per capita income in the area?
- How does local income compare to state and national averages?
- Is there a large percentage of the population living below the poverty line?
- Is the economic base of the community stable all year long, or seasonal?
- Is the economic base stable for the long term or are changes taking place such as industrial layoffs and closings?

Educational factors that may be important in planning include the background of the population and educational participation patterns.

- What is the overall educational attainment of the population?
- Is there evidence that continuing, nonformal education is valued?
- What other educational institutions exist in the area?
- What programming do those institutions provide?
- How is the population organized proportionately with local school districts?

Emotional factors refer to several influences on program planning particularly important in identifying needs and impacting other aspects of programming. Sometimes people express "wants" rather than true "needs". Individuals may not be able to articulate the needs, but the expression of wants may lead to identification of the real need beneath the surface. Sometimes emotional aspects of issues or problems cause them to gain public attention. For instance, the potential location of a hazardous waste disposal site in a local community may create an emotional reaction that provides a "teachable moment" for programs concerning community choices or legal issues of waste disposal. Conversely, emotional factors may inhibit the educator's ability to address particular problems or issues. It may be difficult to offer programming concerning teenage sexuality in a community where religious leaders have taken a prominent stand against sexuality education in schools.

Political influences may be global, national, local, or even within the Extension organizational structure itself. Political influences on program emphasis may be exercised by program funders such as county commissioners, state legislators, or grant providers. Local committee leaders who have been involved in programs over a period of time may perceive they have earned power or authority to govern program directions. A religious community that is outspoken about an issue may influence programming to address related issues.

Personal factors of Extension personnel also influence program direction, though the influence is not often recognized. There are particular areas of interest and expertise an individual brings to his or her position. It is natural to want to use those skills or competencies. Conversely, personnel can thwart programming efforts that are needed but reflect subject competencies in which the professional has little interest or expertise. Extension personnel have the responsibility to ensure that programming is based upon needs rather than simply reflect on what he or she wants to accomplish professionally.

EXPLORING THE STEPS IN PROGRAM PLANNING

The next three sections of this chapter explore the three major components of the program-development process: planning, design and implementation, and evaluation. The key steps in each component are discussed in depth as a practical guide for the inexperienced educator. The first, planning, charts the course for the entire program-development process.

Identifying Program Goals

Program goals are the end or destination toward which programming efforts are directed. Goals may be determined by an individual or by a group. Sometimes goals are established by the organization through strategic and/or long-range planning processes. Goals may also be established jointly with learners in a community or through a partnership of the educator and an individual learner. Boyle (1981) divides continuing educational programs into three categories based upon the primary goal of the program: institutional programs, informational programs, and developmental programs.

Institutional programming focuses on the development of an individual's basic abilities, skills, knowledge, and competencies by teaching content of a discipline. Goals and objectives for institutional programs are developed from the content and needs of the discipline.

Informational programs are characterized by the exchange of information between the educator and the learner. Goals for informational programs are determined by identifying new information that should be disseminated. Both educators and learners may be involved in identifying the general goals of the program. However, the educator and institution play the primary role in structuring goals and objectives.

Developmental programs emphasize helping people with problem-solving or coping strategies. Goals for these programs are determined through a partnership of the educator and learner. Extension programs may include all three types of programs and a variety of means for determining goals. All of these approaches to identifying goals are beneficial.

Conducting Needs Assessment

The Extension educator is continually challenged to identify programs relevant to the concerns of people. It is the educator's responsibility to identify and prioritize the needs of learners in the community or geographical area and to decide how best to allocate resources such as time, money, energy, and personnel to achieve the maximum results.

Needs assessment is the systematic process of analyzing gaps between what learners know and can do, and what they should know and do (Witkin 1984). Needs assessment helps educators improve planning, implementation and evaluation of programs by:

- improving accessibility of programs to a variety of people
- learning more about present conditions in a community
- learning more about specific needs of people in a community
- identifying possibilities to develop new programs or expand existing ones

- assessing public opinion about goals and priorities
- building people's interest in community programs or public decisions

Key questions that needs assessments address include:

- What must be changed or improved?
- What is the real cause of the problem or what issue needs to be addressed?
- Who will be affected by the problem or is involved in the issue?
- What role can education play in solving the problem or impacting the issue?
- Who supports this program?
- What expertise and/or resources exist to address the problem or need?
- Is this need already adequately addressed by others in the community? What other individuals, agencies, or organizations can and will contribute?

Needs assessment involves all people concerned with an educational system: learners, educators, staff, community residents, community and political leaders, concerned citizens, agencies, organizations, and businesses. Involving people is a critical part of the process and must have a specific purpose. Needs assessments may be as broad as an entire organization or as specific as an individual program.

There are many methods for conducting needs assessments. A single technique may be sufficient in one situation, whereas multiple techniques may be necessary in others. The more methods used, the more costly the process in both time and money. Witkin (1984) describes five categories of needs assessment methods: survey methods, social indicators, group processes, futuring methods, and causal analysis.

Survey methods are widely used and have the potential to represent all people in a community. A survey or questionnaire is considered a relatively inexpensive method for collecting information. Surveys may be conducted by telephone, mail, face to face, or through a drop-off/pick-up distribution system. They are often used in combination with other methods. Chapter 8 discusses the development and use of questionnaires in more depth.

Social indicators are the demographic or statistical data that describe the size and characteristics of population groups. Alone, these indicators may not establish need, but in combination with other information, they can provide evidence of need. Social indicators include census data, vital statistics, welfare system input and output data, housing data, government and manufacturing statistics, marriage, death, and health statistics, and other agency or organizational data.

Group processes add another dimension to needs assessment. People who are involved in determining priorities, may gain acceptance for the programs that result. Boyle (1981) suggests three major reasons for involving people in needs assessment:

1. to mobilize support and to overcome resistance
2. to identify solutions for needs and problems
3. to provide an educational experience through participation in planning

Many strategies are used to involve groups in needs assessment. Public hearings are open meetings in which people give testimony concerning needs. Nominal group processes, games, simulations, advisory groups, task forces, and focus groups are all methods that can be used for assessing needs. A number of group processes are described in Chapter 8. A challenge the educator faces in using group processes for needs assessment is accurate representativeness of the community or target population. Rather than using existing groups to assess needs, the educator may want to purposefully identify desirable participants and invite individuals to participate, in order to achieve appropriate representation .

Futuring methods of needs assessment are used for long-range or strategic planning. Futuring methods are proactive and often determined by factors external to the program planner. Anticipating those factors is complex. Several futuring methods include the future wheel, scenario building, decision tree, simulations and games, Delphi, cross-impact analysis, and trend extrapolation.

Causal analysis may be used to look beyond needs to analyze causes. Causal analysis helps the educator avoid making incorrect assumptions or conclusions (Witkin 1984). One way to use causal analysis involves having a planning group examine the causes of problems before identifying possible solutions. For example, the need for more pharmacies to serve the population might be identified. Through causal analysis, you might uncover that there are sufficient pharmacies, but their business hours are limiting accessibility by some people. The causal analysis helped to avoid misinterpretation of the need. Fault-tree analysis is a similar method in which the planner predicts ways a program might fail. Then a group systematically analyzes how to design the program to avoid these pitfalls. A reference that outlines many needs assessment techniques is *Planning and Conducting Needs Assessments: A Practical Guide,* by Witkin and Altschuld.

Overall, needs assessment is an important part of the program-development process. Focusing on critical needs ensures that limited resources of time, money, and expertise are used effectively to help people. The results of needs assessment enable the educator to complete a situational analysis. Each community and audience is composed of a complex set of circumstances. Situational analysis, the description of the setting and circumstances, enables the educator to understand the environment for programming more completely. A checklist for evaluating a situational analysis is provided by Forest and Baker (1994, p. 88):

- ❑ describes the current condition
- ❑ identifies need, problem, opportunity, or emerging issue
- ❑ includes support data and documentation of need
- ❑ includes indicators of severity/scope of need
- ❑ includes benchmark data against which later impact measurements can be compared
- ❑ establishes clear reasons and justifications for program
- ❑ describes primary audience(s), numbers, and geographic locations
- ❑ indicates a gap between "what is" and "what could be"
- ❑ indicates needed research

Setting Program Priorities

To ensure effectiveness of Extension work, the Extension educator must be sure that programs and activities are accomplishing the right ends and addressing the real needs of people. Often many needs are identified without a clear picture of their relative importance or ranking. The process of establishing priorities is ongoing in the Extension educator's daily work. It is easy to become overloaded with programming activities that may not relate to high-priority goals. Extension educators must balance pressures from many sources: the Extension organization itself, local clientele, political leaders, co-workers, society, and individual professional goals.

Priorities must be established in every stage of the program-planning process: defining target audiences, identifying needs, determining methods and strategies, and implementing daily actions to achieve program goals. Priority setting is not just an individual process; it must also involve stakeholders in the program.

Forest and Mulcahy (1976) have identified five important reasons for priority setting. First, priority setting is critical to help direct program action to the changing needs of people. Extension has limited resources in time, money, and personnel to address problems and issues.

Second, priority setting helps us to work proactively to prevent future problems or crises. By having knowledge about the past and research that identifies new solutions to problems, Extension programs can anticipate problems and help individuals and communities take action to avoid crises.

Third, priority setting is important for establishing credibility and accountability with program stakeholders. Extension is increasingly being held accountable for program results by those providing funding: local government, state government, federal government, and granting agencies.

Fourth, the job of the Extension professional is easier if clearcut priorities are established. A common concern of Extension professionals is, "there is so much to do and not enough time to get it done! I can't possibly add one more activity to my calendar." For individuals to maintain physical and emotional wellness, a sense of priorities and limits needs to be established.

Last, setting priorities helps with the process of allocating resources at the local, state, and federal level. If priorities are determined, limited resources can be used more effectively. If resources are distributed across too many programs, none of the programs will be as effective as they could be. Fewer, more focused programs generally have greater impact on the targeted audiences. In addition, the Extension professional can focus his or her work in a more effective manner.

Successful priority setting depends upon implementation of several critical steps:

Step 1: Assessing the present situation by identifying all the ongoing programs and activities to which an individual is currently committed.

Step 2: Identifying possible priorities from needs assessment and organizational sources.

Step 3: Weighing the importance of priorities to compare all programs and activities identified in steps 1 and 2.

Step 4: Considering consequences of acting on priorities: What will be the impact if a program is not continued? What will be the impact on Extension personnel if another program responsibility is added to an already full schedule?

Step 5: Taking action on priorities. Actions may include referring a program to others, clarifying responsibilities among person nel, shifting emphasis from low-priority to higher-priority programs, reviewing existing programs for change, and establishing a time line for action (Forest and Mulcahy 1976).

Identifying Target Audiences and Capabilities

Once the needs have been identified and program priorities determined, the next step in the program-planning process is to identify target audiences and the means to appropriately address their needs. Why not design generic programs for use with anyone and everyone? People have different needs and wants concerning programming based upon many factors. Age, educational level, socio-economic status, language, disability, ethnicity, cultural patterns, participation patterns, and occupation are a few of these factors. Educational programs will more effectively address the needs of individuals when designed with the specific audience in mind. For example, a nutrition program focusing on protein alternatives in the diet would differ for a vegetarian audience and an audience that consumes meat. A program about business management of farm enterprises will be more meaningful to farmers when the type of business is targeted, such as Management for Dairy Producers or Managing Your Commercial Landscaping Business. A program designed for eight-year-old children will probably not appeal to teenagers.

Several key questions are helpful in targeting audiences in program planning.

1. Who are the people affected by this problem or issue? What different audiences need to be addressed?

   Examples:

   A need has been identified for programming concerning waste disposal alternatives. Possible audiences to target with different dimensions of the problem are:

   - youth: value of recycling and developing recycling habits
   - homeowners: recycling alternatives available in the community
   - businesses: how to recycle in the work environment
   - farmers: opportunities to use recycled newspaper as animal bedding
   - city officials: waste management options for communities, i.e. landfills, incinerators, curbside recycling, etc.

2. What are the specific characteristics of the audience that impact how to plan and deliver programs?

   Examples:

   - Dual wage-earner parents of preschoolers may not be available for daytime parenting programs but would participate in an evening or work-site program.
   - Limited-income families may not be able to afford to send several children to an overnight camp but a day camp may be more economical.
   - A training program for Hispanic farm laborers about safe handling of pesticides will be more effective if conducted in Spanish rather than English.

3. What skills/strengths might this target audience have that programming can further develop and/or enhance?

   Examples:

   - School enrichment programming on teen decision making could capitalize on their interest in purchasing their first automobile.
   - A parenting program for single parents could include participants' success strategies for building support networks.
   - A minimum tillage field day could include a panel discussion of farmers who are already using these practices to share benefits and barriers.

4. What social, cultural, economic, or environmental changes should occur in this situation? Are these changes possible and realistic for this audience?

   Examples:

   - Is an international exchange program to develop youth cul-

tural awareness realistic if no plans are included to obtain financial support for interested youth?
- Is it reasonable to expect that a three-week series of classes on managing your food dollar would result in low-income families being able to get by without food stamps?
- Is it feasible to teach farmers to recalibrate sprayers in one hands-on workshop?

5. What timing factors are important to consider for various groups?

Examples:

- Would grain farmers attend an estate planning workshop during planting season?
- Would attendance be affected at a teen leadership camp planned the same weekend as many local high school graduation ceremonies?
- Would homeowners be more interested in a workshop to learn to prune fruit trees at the time their trees need to be pruned?

These may appear to be very simple, logical factors to consider. Yet many times Extension educators make errors that impact program participation and effectiveness. Considering these factors during the planning phase of programming can help educators avoid discouragement, disappointment, and loss of time and effort.

Writing Program Objectives

Precise statements defining what program participants should do upon completion of the program is an important element of planning. Objectives should be prepared to reflect the levels of development program participants are striving toward. One program may have several objectives at different levels. Precise program objectives serve as the criteria on which to base evaluation. Objectives may target participation of people, reactions of people, knowledge gain, attitude change, skill development, changed aspirations, change in practices used, and ideally, long-term behavioral change. Effective educators include representatives of the target audience when developing realistic and meaningful program objectives. Additional appropriate objectives may become apparent during the process of program implementation and can be added at that time.

Bennett's Hierarchy (1975) is a model that has been used frequently in Extension to describe different levels of objectives and program accomplishments, see figure 5–3. Each level of the hierarchy indicates a criterion for developing program objectives and for measuring program results. Extension programs usually have several objectives that may include several levels of the hierarchy. The first two levels provide little to no information about how the program participant has benefitted, but may be pertinent for Extension personnel. *Inputs* and *Activities* are appropriate for setting objectives pertaining to allocating resources to programs planning particular educational activities. For example, a level 1 objective

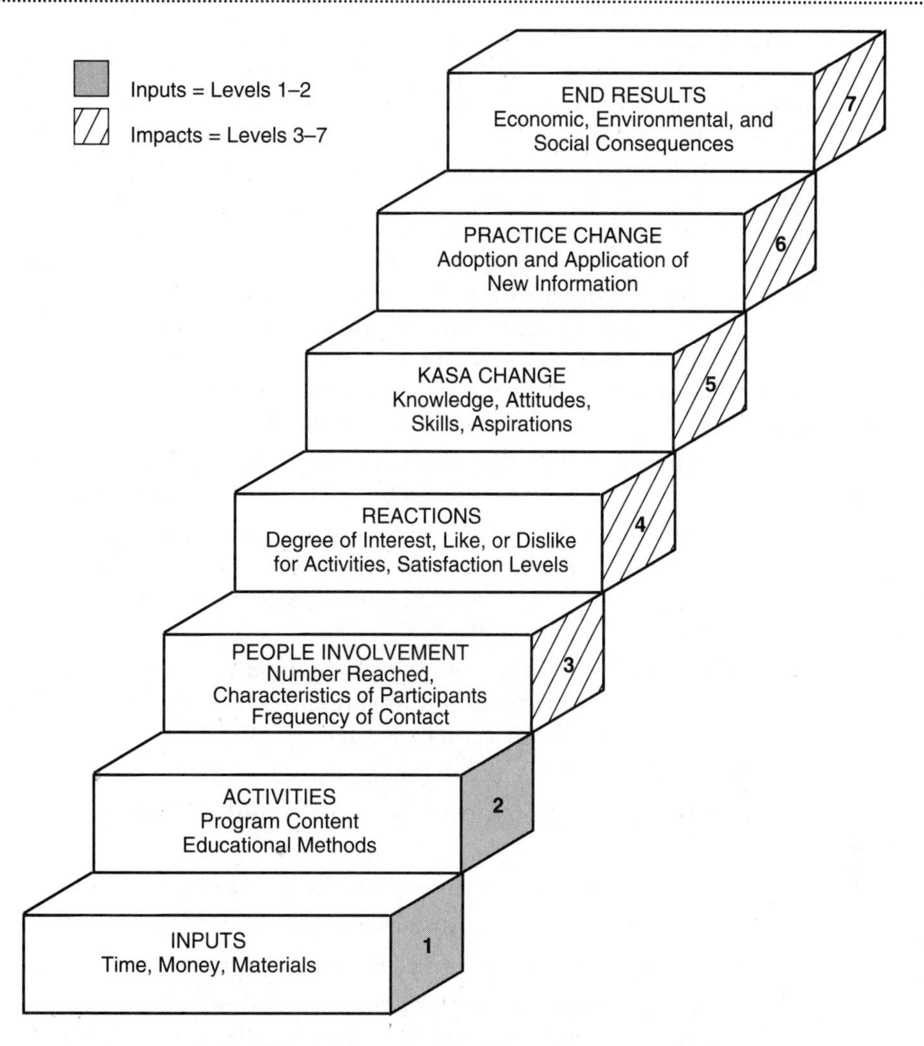

Figure 5–3 ***Bennett's Hierarchy. Adapted from Bennett, C.F. (March, April, 1975). Up the Hierarchy.*** **Journal of Extension** ***13 (1):7–12.***

might be that after one year of a program, all costs for program materials would be generated through registration fees. An example of a level 2 objective would be the intention to conduct six county workshops during a particular time frame.

As the hierarchy progresses the evidence of program impact upon people becomes stronger. An objective written at level 3 (*People Involvement*) would target the number of people one planned to reach with a particular activity. Level 4, *Reactions*, identifies the type of reaction one intends to elicit from program participants, such as people demonstrating interest through subscribing to a newsletter. Level 5, commonly called *KASA*, refers to the knowledge, attitudes, skills, and aspirations generated in participants as a result of the program. Objectives at this level may target increases in knowledge about a subject, improvement in ability to complete a particular task, change in attitude about a particular idea or practice, and goal setting by participants concerning future action.

Levels 6 and 7 (*Practice Change* and *End Results*) are at the highest levels of the hierarchy. As one might anticipate, achieving program objectives and impact at these levels is more difficult than at the lower levels. At the *Practice Change* level, one targets changes in participant behavior that involve application and adoption of ideas or practices. *End Results* targets the long-range outcomes or effects of the program over time.

Using the ABCD's to write objectives can be a helpful tool:

A-Audience: Describe the audience, the individual learners who are to acquire the new capability.

B-Behavior: Describe the behavior identified and name what the learner should be able to do after the program.

C-Conditions: Describe the conditions that will prevail while the learner carries out the objective and what the learner will or will not have accomplished at the time of assessment.

D-Degree: Describe how well the behavior must be performed or the criteria and standards that will be used to evaluate performance of learners.

Objectives can be evaluated using the checklist below:

- ❑ Does the objective indicate what value will result from the program?
- ❑ Does the objective indicate who will benefit?
- ❑ Is this objective clear and meaningful?
- ❑ Can objectives be achieved through educational programs?
- ❑ Does the objective include what the target audience is expected to know and to do as a result of the program?
- ❑ Is the objective specific enough to be measured?
- ❑ Are there preset standards that can be used to determine how well the objectives have been met?
- ❑ Does the language in the objective make sense to those people who will be using and reading the objective?
- ❑ Does the objective contain all major points needed to give staff focus for the program and to communicate to others about the focus of the program?

PROGRAM DESIGN AND IMPLEMENTATION

After identifying community needs, establishing priorities, targeting audiences, and writing objectives for programs, it is time to design the program-implementation plan. There are several components to this phase of planning: selecting and/or developing educational content, selecting program delivery methods, identifying or developing resource materials, and constructing a time line for implementation of the program.

Selecting and/or Developing Educational Content

The key consideration is what content or process is necessary to solve the problem or resolve the issue. Content may differ to serve the needs of each targeted audience in order to achieve the desired results. Content needs to be sequenced logically for effective learning to occur. There are many resources available to the Extension educator for developing and selecting program content. The Extension system is built with an internal network of content-support specialists. This network may include other county agents, district or regional program specialists, and state program specialists. These individuals work individually and in teams to develop and/or identify existing content resources for programming. Standardized curricula exist for some types of programs. For others, new approaches may be necessary. In addition to the internal resources of Extension, the educator may seek other experts outside the system. Local coalitions may work together to develop program content or to adapt an existing curriculum.

Selecting Program Delivery Methods

When planning program delivery methods, several questions need to be considered.

- What educational activities/learning experiences are most appropriate for the audience and content?
- Are the program delivery methods selected appropriate for accomplishing the identified objectives?
- What is the most logical sequence of activities and/or methods?
- Do the learning activities build logically on other learning experiences?
- What routine or ongoing activities and events can be used to enhance this program? Examples include newsletters, news columns, radio programs, television programs, existing clubs, advisor training, or other organizational activities.
- What events and activities will be held?

Chapters 6 and 7 cover the teaching-learning process and Extension teaching methods in depth.

Selecting and/or Developing Resource Materials

The national Extension organization has a wealth of resource materials for programming. Subject-matter specialists in each state are an important resource to Extension agents for current program materials. Electronic databases such as PENPAGES are emerging as a means of sharing programming resources across the world.

When selecting and/or developing programmatic resources, consider how the materials support the content of the program.

1. What materials are needed: bulletins, fact sheets, videotapes, public service announcements, graphics, games, experiential learning activities? Are the materials appropriate for the identified objectives?
2. Do the materials address the needs of the specific target audience? For example, if an individual with a visual disability requires a tape-recorded publication, how can this be obtained? Are written materials suitable for the reading level of the intended audience? Do visual materials contain people to whom the target audience can relate culturally?
3. Are the needed materials available? From whom or where? Are there costs involved? How will the cost of materials be covered? How long will it take to obtain the materials?
4. If materials are not currently available, who will develop them? What time frame will be needed to develop the necessary materials? Who has the expertise to assist in this area? Will you work independently or with a team? Do others who will assist understand the objectives of the program in which they are involved?

Constructing a Time Line for Implementation

Time lines are an effective tool with which the Extension educator can outline and track tasks necessary to plan, implement, and evaluate a program. Time lines can also help volunteers who are assisting with various parts of the program-planning process. To be effective, time lines need to list specific activities to be accomplished as well as when each activity needs to be done. When developing a time line, it is helpful to begin with the target date for completion of a project or program and work backward to ensure adequate lead time for completion of each step. Many experienced Extension professionals develop time lines for projects and programs that are repeated regularly.

PROGRAM EVALUATION

Although Chapter 8 is devoted to the topic of evaluation, it is appropriate to explore the importance of including planning for evaluation in the program-planning process because evaluation is essential to successful programming in several ways. First, needs assessment—a form of evaluation—provides critical information about the community for formulation of program goals and objectives. Second, feedback gathered by evaluation throughout the program planning and implementation process, allows for adjustments or mid-course corrections. Third, evaluation of teaching during the program implementation phase provides educators with sound information on which to improve their effectiveness as teachers. Finally, evaluation provides documentation of program outcomes and end results. This information is critical for documenting program impact, making changes for future programs, and identifying additional goals and objectives for future programming. In addition, Extension educators are increasingly required to provide evaluation documentation to administrators and program funding sources.

Evaluation is frequently left to last for consideration in program planning. What occurs then is that educators move on to the implementation of another program and neglect the evaluation phase of the previous program. By planning for evaluation during the program-development process, all the steps necessary to complete the entire process (including evaluation) can be included in programming time lines.

INVOLVING PEOPLE IN PROGRAM DEVELOPMENT

Program development in Extension is based upon a philosophy of social change—mobilizing and developing human resources in ways that enable programs to be implemented most effectively (Boyle 1981). Boyle describes three major reasons to involve people in program development:

1. To reach more accurate decisions about relevant needs and opportunities for programming
2. To speed the change process through diffusion and legitimization of programs
3. To involve people in learning experiences that prepare them for active leadership in the change process.

The educator needs to purposefully include people in the program-development process by creating situations that capture their attention and interest. This can include a wide range of activities, such as those described in the needs assessment section of this chapter. Committees and task forces are commonly used in Extension to involve people in program planning. However, before planning how to involve people in program development, it is critical to determine the purpose for their involvement.

Committees and Program Development

Committees are commonly used in organizations within democratically governed nations as a way to ensure input from the public. Besides providing input for program planning, committee members contribute time, energy, and expertise to the implementation of programs. Members are often selected to represent people in the community who are similar in demographic characteristics or interests. It is essential that a committee have a purpose for being. Without a well-defined purpose, a committee may create its own purpose, which could be in conflict with the organizational program goals and directions, or the committee members may become frustrated and not function effectively. Some of the possible functions of committees include:

- identifying and prioritizing needs, problems, and resources
- legitimizing program decisions made by the educator or the organization
- identifying alternatives and solutions
- creating awareness and promoting programs among the public
- assisting with specific aspects of program implementation and evaluation
- assist with financial support of programs

Figure 5–4 outlines some key functions of committees and possible activities in which they may be involved.

Two major types of committees are commonly used in program planning: standing committees and ad hoc or special-purpose committees. Standing committees handle ongoing responsibilities in program planning. Ad hoc or special-purpose committees are formed to complete a particular task or responsibility and are generally short-term in duration.

There are several advantages to using a committee as compared to a general group meeting for program development:

- **Size:** Committees can be more effective than general group meetings because they are smaller in size and each member has greater opportunity to participate.
- **Informality:** Procedures can be kept informal with few rules.
- **Composition:** If committee members are identified based upon interest, participation will be enhanced.
- **Communication:** Controversial or confidential subjects can be discussed more easily in a small group.
- **Scheduling:** Smaller groups are easier to coordinate with scheduling
- **Function:** Committees can operate more efficiently than a large group because members learn to work together over time.

FUNCTIONS OF COMMITTEES	COMMITTEE ACTIVITIES
Fact Finding	Collect specific kinds of data
Advisory	Advise on policy, technical matters, program needs, procedures, etc.
Program	Plan specific events
Membership	Solicit new members, maintain records, collect dues
Public Relations	Inform public about purpose, goals, activities, events, and accomplishments
Coordinating	Coordinate activities of two or more organizations, committees or groups

Figure 5–4 *Functions of Committees. Adapted from:* **Committees . . . A Key to Group Leadership.** *(June 1980). North Central Regional Extension Publication No. 18, p.5.*

The purpose of the committee determines what characteristics are required of potential members. Generally, the educator seeks people who are interested in community improvement, have knowledge about the situation, are willing to invest time, and have the skills or experience needed to work in a group setting. Other special characteristics may be needed depending upon the purpose. For a task force to examine business and retention needs in a community, volunteers may need to be familiar with local business and industry leaders and have confidence in making contact with these leaders. For a committee to conduct a 4-H fundraiser, members may need good organization skills and creative ideas for making money.

Committees have been shown to be more effective if the members understand the objectives of their involvement, the means they have available to accomplish their objectives, the power they have to use those means, and the importance of planning (Lacy, in Boyle 1981). It is the educator's role to ensure that members have the appropriate orientation and resources to accomplish their purpose. Goals and objectives of the committee and roles of the members need to be discussed when a committee is formed. If a committee is ongoing, with rotating members, plans should be made to orient new members as they join the group.

The Extension professional can use several strategies to help committees function smoothly. As committees are appointed, the purpose, nature, and composition of the committee should be clarified. Provide committee members the purpose, duties and responsibilities, and duration of committee appointment in writing. Help the group determine an identifiable name if one does not currently exist. Committee members need to know the time commitment expected from them as well as the length of their term of service on the committee. Stating the function of the committee and any boundaries of responsibilities and activities is important. Otherwise committees can take control beyond their sanctioned area of responsibility. Committee members need to know what resources are available: money, volunteer staff, office equipment, materials, etc. The structure and makeup of the committee should be clearly defined, including whether officers are elected or appointed.

The Extension professional should facilitate communication among committee members by providing a list of names, addresses, phone numbers, and other pertinent information. Meetings should be scheduled well in advance and with the input of members concerning convenient dates and times. If a committee functions with subcommittees, indicate in writing the specific duties or assignments for each subcommittee. Ensure that committee records of membership lists, agendas, and meeting minutes are maintained by a secretary or other member. Keep copies for your records.

Maximizing Impact with Coalitions

A *coalition* is defined as a grouping of interest groups who are committed to achieving a common goal (Bacharach and Lawler 1981). The work of coalitions is often described as cooperation among groups who have agreed to communicate and coordinate actions (Groennins, Kelley, and Leiserson). Contrasted to a committee, members of a coalition usually represent a particular organization and participate in joint planning on behalf of that organization. Coalitions are being emphasized as an effective way for agencies and organizations to work together to eliminate duplication of effort as well as a means to pool financial and human resources to accomplish shared goals.

Coalitions are also used in the program-development process. The role of the Extension educator may be to organize and facilitate a coalition concerning a particular need or problem or it may be to serve as a member of a coalition representing Extension programming. Coalitions may combine public organizations, private organizations, or a mixture. Organizations may represent a single community or problem, or they may represent different levels of government. How do you decide if a coalition is needed? Sometimes coalitions are mandated by law—such as a coalition to address nutrition education of parents on public assistance. Limited resources may lead to the formation of a coalition to maximize money, time, or expertise available. Or a coalition may be formed because there is adequate time and interest among parties to develop a different approach to a problem or activity.

As coalitions are formed, several key questions should be considered to ensure positive outcomes.

- Is each organization committed to solving the problem?
- Is each organization committed to coordinate its work to solve the problem?
- Does each organization believe that every other organization has the right to be involved?
- Do the benefits outweigh the costs for each organization? Consider whether coordination will
  - detract from the organizational image
  - save or cost time
  - save or cost money
  - result in better or worse service to clientele
  - result in a better or poorer quality product
  - promote or suppress innovation or new programs

To be successful in working with coalitions, the Extension professional should emphasize the goal of the group to clarify purpose and communication. The group coordinator may need help from a "legitimizer" to get the group going. This could be the organization providing funding, a political leader with power to establish the group, or a community leader who positively influences others. Then it takes time to establish trust among the coalition members by following through with agreed upon commitments. Group members must agree upon how to share costs, credit, visibility, and funding so all members feel valued and important to the effort. If problems occur, the group may need to refocus priorities and redefine who the essential partners are. To avoid pitfalls, it is important to be explicit about work arrangements and anticipate some frustration. Despite the difficulties, coalitions can provide a new and exciting way for organizations to accomplish program goals.

PROGRAM MANAGEMENT STRATEGIES

Experienced Extension personnel utilize a number of strategies to assist them in managing the multiple tasks involved in program planning. A *task analysis* worksheet is helpful the first time an Extension educator completes a particular program, see figure 5-5. The task analysis should include activities that must be accomplished, when each activity needs to be done, who is responsible for the activity, how the activity will be accomplished, what resources are available to do the program, and a general plan of action. A task analysis worksheet can be constructed as the program-planning process occurs.

Reminder lists are useful for many dimensions of program development. Lists for the programmer might include materials needed, people to contact, or action steps. The list may be for your use or for use by volunteers, support staff, youth leaders, or other professionals.

Notebooks for professionals or volunteers help to organize materials for programming. A table of contents, besides listing the items contained in the notebook, may help organize tasks to be accomplished. Dividers may be used to group information into categories. Extension professionals may develop notebooks to use with 4-H advisor orientation, camp counselor training, advisory committee member orientation, Master Gardener training, etc. Notebooks are also useful to organize materials for teaching. Teaching outlines, overhead transparency masters, notes, sample handout materials, and other resources may be arranged by topic area.

Worksheets can be helpful in developing action plans for new programs or to assist others in outlining activities. Worksheets may cue the

GENERAL TASKS AND SUB-TASKS	TASK BEGIN/END DATE	NAME OF PERSON(S) WHO WILL COMPLETE	ESTIMATED PERSONNEL COST	OTHER RESOURCES NEEDED	ESTIMATED COST FOR OTHER RESOURCES
TASK:					
TASK:					
TASK:					
TASK:					
TASK:					

Figure 5-5 *Task Analysis Worksheet. Source: Adapted from:* **Committees . . . A Key to Group Leadership** *(June 1980). North Central Regional Extension Publication. No. 18, p. 5. Columbus, OH: The Ohio State University, Cooperative Extension Service.*

user to categories for consideration in planning, such as a budget planning worksheet. The programmer needs to consider costs of personnel, travel, materials and supplies, communications and marketing, indirect costs, etc. A worksheet for 4-H advisors may help them plan the club meetings and member responsibilities for the coming year.

Checklists help track activities or tasks that need to be completed as well as those that have been accomplished. They are useful for individuals or for teams. Extension professionals may provide checklists to volunteers or committee members to help them organize work to be done.

There are many more program management strategies that assist Extension personnel in keeping the program planning process on track. The key is to identify several that work well for the individual and use them regularly to assist in program action.

COMMUNICATING PROGRAM PLANS

Once program plans are in place, Extension personnel have a responsibility to communicate them to others in the Cooperative Extension System. An administrative reporting system designed to communicate program-plans and outcomes is called Plan of Work (POW) and Report of Results (ROR). Most state Extension systems utilize some type of individual or team Plan of Work and Report of Results. The Cooperative Extension System at the federal level requires each state to develop a Four-Year Plan of Work that is updated annually and includes annual narrative and statistical reports of accomplishment. Generally, POW and ROR systems include several basic elements: a statement of the situation or need, program objectives, identification of target audiences, program delivery methods, and strategies for evaluation. Each state has a system for using these reports to summarize statewide programming efforts and impacts to share with administrative leaders, legislators, and key leaders. POW and ROR systems communicate meaningful information only when the process used to plan, implement, and evaluate the programs is sound.

NEW DIRECTIONS IN EXTENSION PROGRAM PLANNING: ISSUES PROGRAMMING

As described in Chapter 4, Extension programs have historically been based in the disciplines of agriculture and natural resources, home economics, community development, and 4-H. Inherent in that programming structure was a focus on particular populations or groups in a community: rural families, agricultural producers, rural community leaders, rural youth or youth enrolled in 4-H clubs, and Family and Community Education group members (formerly known as Extension Homemakers). The traditional approach to program planning included meeting with representatives of these respective groups and asking, "what programming do you need or want from Extension this year?" Sometimes groups focused on the method of delivery, like an annual tour, rather than the need for the content of the program. As demographics of the United States shifted and fewer people were part of the traditional rural audiences, Extension refocused on a wider cross section of the population in communities. This has resulted in a broad list of needs that may require interdisciplinary approaches to develop programming that reaches beyond the knowledge base of agriculture and natural resources, home economics, community development, and 4-H.

Since 1985, the National Initiative Coordinating Committee has worked to develop a framework for Extension programming across the United States that is proactive, relevant, and integrated not only among the disciplines in agriculture, natural resources, and home economics, but also in areas that have not previously been involved in Extension programming. The organizational structure for programming is determined by identified needs and the appropriate discipline base and professionals involved in the programming process. The underlying purpose of this redirection is to provide relevant research-based education to improve people's lives—a continuing commitment to the land-grant mission.

The effect upon the delivery or implementation of such a mission is that the problem or issue defines the audience, the method of implementation, and the combination of personnel and resources to address the issue (Dalgaard et al. 1988). This approach is known as **issue-based programming.** *Issues* are defined as matters of wide public concern arising out of complex human problems (Dalgaard et al. 1988).

An issues-based approach to educational programming has several benefits. Program development is based upon broad-based community needs assessments. Community issues are identified and become the framework for developing programs. The educational process assists people by examining multiple perspectives concerning an issue and providing research-based information to assist individuals in forming a perspective. In addition, the issues programming approach emphasizes the importance of the social action process; it helps people learn how to take action to impact issues and change a situation themselves. The information sources used to assist people could reflect the expertise available throughout a university. Since issues change over time, educational program directions change to be relevant to the community. Issues programming focuses upon being proactive rather than reactive, helping learners focus upon the future and assisting them to develop change adoption strategies (Kast 1980).

Issues are complex sets of problems. The audiences to be addressed, the methods of program delivery, and the discipline base for the program cannot be predetermined by the Extension educator. This approach emphasizes the need for teams and networks to collaborate on the planning, implementation, and evaluation of programs (Richardson 1991). Figure 5-6 illustrates the program-planning process once issues are identified.

The critical difference between discipline-based programming and issues programming is the goal: the goal of discipline-based programming

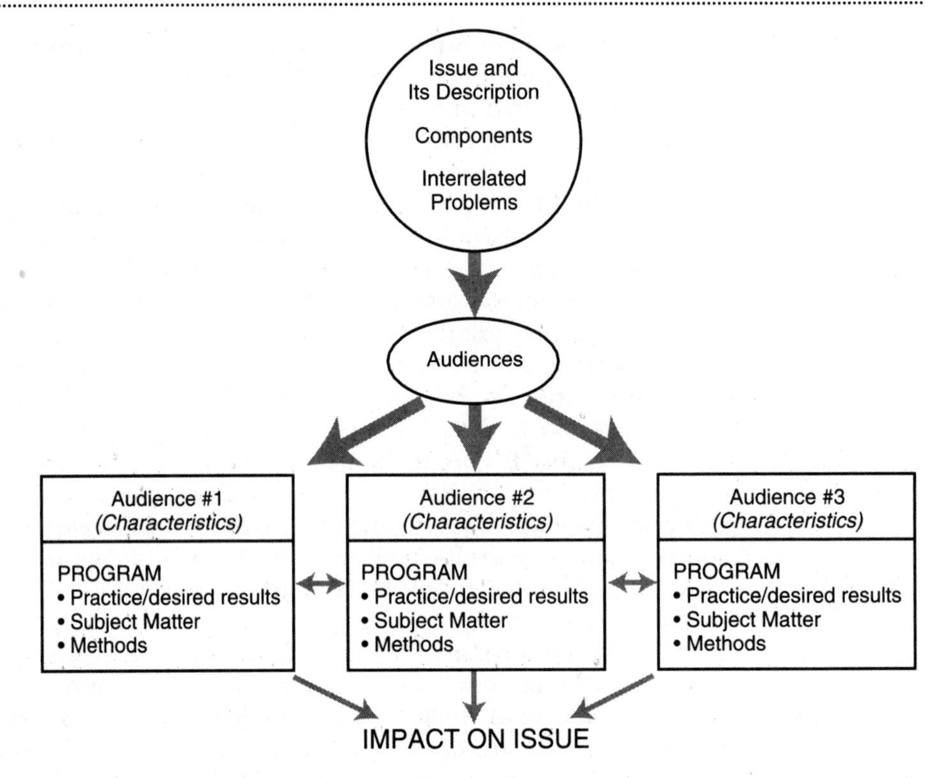

Figure 5-6 ***Creating Programs to Address Issues. Source: Richardson, B. (1991)*** **An Introduction to Issue-Based Programming.** ***Texas Agricultural Extension Service, The Texas A&M University, College Station, Texas. p.4.***

is knowledge transfer; the goal of issue-based programming is solving problems through education. Issues programming represents the practice of sound program development. The Extension System often describes disciplinary programs as *Base Programs* and refers to issues programming as *National Initiatives.*

ETHICAL ISSUES IN PROGRAM PLANNING

In an era when more sources of funding for educational programming require specific goals, objectives, target audiences, or methodologies from programming, the subject of ethics surfaces. A research study concerning criteria for research problem choices among agricultural scientists found that personal enjoyment ranked higher than public need (Busch and Lacy 1983). Scholl (1989) conducted a similar study with Extension home economists to identify what influences their program-planning priorities. The study found that the educators used both formal and informal needs assessment strategies to determine program direction. Personal philosophy, educational background, and job location also impacted planning.

Although the educator is central to the program-planning process, the role of the learner or potential learners is critical. Enough concerns have surfaced about the lack of learner involvement in program planning to warrant attention to a number of ethical issues. Sork (in Brockett, 1988) cautions adult educators about nine ethical issues in program planning:

1. Responding to "felt" or "expressed" needs of adult learners
   - Does conducting a needs assessment signal that all needs will be addressed, giving learners false hopes?
   - Whose values are reflected in how needs are addressed?
2. Basing a program on a need not acknowledged by the learner
   - Is the educator making inappropriate assumptions about what the learner needs and designing programs that address needs not acknowledged by learners?
3. Basing program planning on learning "deficiencies" of adults
   - What are the long-term consequences of basing programs on deficiencies rather than capabilities?
   - What is the impact on the learner's self-esteem?
   - Is the educator imposing his or her value of what is a deficiency?
4. Claiming that capabilities will be developed by learners who participate in a program
   - Does specifying anticipated outcomes misrepresent the complexity of learning?
   - Is it acceptable to describe the structure of a program without specifying the objectives?
5. Designing programs in which participation is compulsory
   - Is forcing an adult to participate in education immoral?
   - In what situations is it justified?
   - Does participation always guarantee improved practice?
6. Maintaining confidentiality of information
   - What information collected by an educator should be public?
   - What information should be private?
   - Is it justifiable to release sensitive information gathered during planning?
7. Selecting instructional and other resources
   - How much energy should planners be expected to expend seeking out resources such as funding?
   - Should planners be expected to reveal the criteria they use to make judgments about program resources such as speakers and materials?
8. Deciding who will be involved in the planning process
   - Is it morally acceptable for a program to be developed with only a program planner and content expert involved?
   - Are there forms of "token" involvement of learners that are improper?

- Is involvement of the learner and others in the planning process a means for the planner to avoid accountability by placing blame or scapegoating?

9. Determining fees for the program
   - Is it ethical to set a program fee knowing that those who might benefit most will be unable to pay?
   - Is it ethical to charge more for some programs to subsidize other programs for those unable to pay?
   - Are pricing policies used to restrict some program participants intentionally?

There are no right or wrong answers to address these issues. They are matters that every Extension professional or adult educator needs to think about and discuss with others. These and other issues have led some groups of professionals to develop codes of ethics for professional behavior. In 1991, the Board of Directors of the Coalition of Adult Education Organizations (CAEO), an organization representing twenty-six national adult and continuing education organizations adopted a "Bill of Rights for the Adult Learner." Several of the "rights" specifically address dimensions of program planning:

> *The right to participate or be appropriately represented in planning or selecting learning activities in which the learner is to be engaged.*
>
> *The right to a learning environment suitable for adults to include appropriate instructional materials, equipment, media, and facilities.*
>
> (CAEO February 21, 1991)

Programming Pitfalls

The Extension Committee on Policy Program Development Task Force studied the program-planning practices of Extension educators to determine how closely they follow theoretical program-development processes. The Task Force identified nine "programming pitfalls" or actions that Extension professionals fail to practice that, if followed, would lead to stronger programs:

- using planning groups that aren't representative of the people for whom the program is intended
- working alone on programs rather than cooperating with other professionals and organizations who could contribute
- focusing on "doing" the program rather than "planning"
- failing to address all the steps in the program-planning process
- failing to prioritize needs or problems to provide clear program direction
- conducting programming that doesn't relate to local needs or problems
- allowing interests of the professional, rather than needs of people, to dominate programs

- viewing program plans as administrative rather than useful
- not evaluating programs (Oliver 1977)

These "programming pitfalls" serve to remind new and experienced professionals of how important program planning is to the desired outcome of successful programming.

SUMMARY

Program planning is critical to the success of Extension programs. Knowledge and skill in practicing program planning are essential job functions for Extension personnel. Though many models of program development exist, they all include three basic elements: planning, design and implementation, and evaluation and accountability. Successful planning involves many people in a systematic process to ensure that the resulting program addresses critical needs of people using methods appropriate for the intended audience.

DISCUSSION QUESTIONS

1. What are the key reasons for setting programming priorities?
2. How can the programmer involve people in the community in the program-planning process?
3. Identify the types of information useful in program planning and where an Extension educator might obtain it.
4. What are the principle barriers to effective program planning?

REFERENCES

Bacharach, S., and Lawler, E. 1981. *Power and Politics in Organizations.* San Francisco, CA: Jossey-Bass Publishers.

Bennett, C. 1975. Up the hierarchy. *Journal of Extension* March/April: Vol. XIII, 7–12.

Boone, E. 1985. *Developing programs in adult education.* Englewood Cliffs, NJ: Prentice-Hall.

Boyle, P. 1981. *Planning better programs.* New York, NY: McGraw-Hill.

Busch, L., and Lacy, W. 1983. *Science, agriculture and the politics of research.* Boulder, CO: Westview Press.

Butler, L., and Howell, R. 1980. Coping with growth, Community Needs Assessment Techniques. Western Regional Educational Publication #44. Corvallis, OR: Western Rural Development Center, Oregon State University.

Coalition of Adult Education Organizations. 1991. *Bill of rights for the adult learner.* Statement of Rights for adult learners affirmed by the Board of Directors of the Coalition of Adult Education Organizations on February 21, 1991, Washington, D.C..

Conklin, N., and Spiegel, M. 1992. *A systems approach to program development and evaluation.* Unpublished manuscript. Columbus, OH: Ohio State University Extension.

Dalgaard, K. et al. 1988. *Issues programming in Extension.* St. Paul, MN: Extension Service-USDA, Extension Committee on Policy, and the Minnesota Extension Service.

Extension Committee on Organization and Policy Program Development Ad Hoc Committee. 1974. *Extension program development and its relationship to Extension management information systems.* Ames, IA: Iowa State University.

Forest, L., and Baker, H. 1994. The program planning process. In *Extension Handbook*, Blackburn, D., ed. Toronto, Canada: Thompson Educational Publishing.

Forest, L., and Mulcahy, S. 1976. *First things first: a handbook of priority setting in Extension.* Madison, WI: University of Wisconsin Extension.

Groenning, S., Kelley, E., and Leiserson, M. 1970. *The Study of Coalition Behavior.* New York, NY: Holt, Reinhart, and Winston, Inc.

Kast, F. 1980. Scanning the future environment: social indicators. *California Management Review* 23(1):22–31.

Mayeske, G. 1994. *Life cycle program management and evaluation: an heuristic approach.* Part 1. Washington, D.C.: Extension Service, USDA.

Mustian, D., Liles, R., and Pettit, J. 1988. *Working with Our Publics, The Program Planning Process*, Vol. 2. North Carolina Agricultural Extension Service and the Department of Adult and Community College Education, North Carolina State University, Raleigh, NC.

Oliver, C. 1977. Toward Better Program Development. *Journal of Extension* 15:18–24.

Richardson, B. 1991. *An introduction to issue-based programming.* College Station: TX. Texas Agricultural Extension Service, Texas A&M University.

Scholl, J. 1989. Influences on program planning. *Journal of Extension*:18–20.

Sork, T. 1988. Ethical Issues in Program Planning. In *Ethical Issues in Adult Education*, Brockett, R. ed. New York, NY: Teachers College, Columbia University Press.

Sork, T., and Buskey, J. 1986. A descriptive and evaluative analysis of program planning literature, 1950–1983. *Adult Education Quarterly* 26(2):86–96.

Spiegel, M., and Leeds, C. 1992. *The answers to program evaluation, a workbook.* Columbus, OH: Ohio Cooperative Extension Service, The Ohio State University.

Witkin, B. 1984. *Assessing needs in educational and social programs.* 1st ed. San Francisco, CA: Jossey-Bass.

Witkin, B. R., and Altschuld, J. R. 1995. *Planning and conducting needs assessments, a practical guide.* London, England: Sage Publications.

Worthen, B., and Sanders, J. 1987. *Educational evaluation, alternative approaches and practical guidelines.* White Plains, NY: Longman.

CHAPTER

"To raise new questions, new possibilities, to regard old problems from a new angle, requires creative imagination."
–ALBERT EINSTEIN

The Teaching-Learning Process

Key Concepts

- Extension's role in the educational process
- The nature of adult education
- The teaching-learning interaction
- Motivation for participation in education

THE CONCEPT OF EDUCATION

The word *education* brings to mind many different images, yet the most deeply rooted is the one associated with young people in a formal school setting. Webster (1994) reinforces this image with the definition of education as "the process of training and developing the knowledge, skill, mind, character, etc., especially by formal schooling, teaching, and training." But Cremin (cited in Darkenwald and Merriam 1982, p. 2) defines education as "the deliberate, systematic, and sustained effort to transmit, evoke, or acquire knowledge, attitudes, skills, or values as well as any outcomes of those efforts." Taken in this much broader context, it is clear that education can and does occur in a variety of settings and through a variety of activities and events. Although schools and universities are important institutions that educate, they most certainly are not the only ones or necessarily the most important ones. Education is an ongoing process that continues throughout life. During a life span, individuals will engage in educational opportunities through a variety of institutions and experiences. Education is an integral part of living.

Formal versus Nonformal Education

In the United States, school attendance is compulsory until age sixteen. Almost all modern societies provide some type of formalized education system at the secondary and post-secondary level. A description of these formal systems usually produces a similar list of characteristics: compulsory attendance, disciplinary techniques and policies, and a defined curriculum. The teacher assumes control for determining needs, developing objectives, and selecting the delivery methods and techniques to be used. Evaluation strategies are almost always conducted in the form of testing and written assignments. Formal education systems serve a vital and necessary role, but they are only one approach.

Nonformal education, on the other hand, can involve so many different activities and approaches that it is difficult to identify a single definition. Coombs (1973) describes nonformal education as:

> *. . . any organized educational activity outside the established formal system—whether operating separately or as an important feature of some broader activity—that is intended to serve some identifiable learning clientele and learning objectives.*

Summaries of the literature on nonformal education by Etling (1975) and Khan (1989) identified the following six dimensions of such programs:

1. A **learner-centered approach** focuses the emphasis of the program on the learning activities and not the teaching that takes place. The learner is actively engaged in the educational process and assists in developing objectives and determining program content and methodology.
2. Curriculum options demonstrate **variety and flexibility**. An important aspect of nonformal education is that it is practical, flexible, and based on the needs of the participants.
3. Nonformal education is based on **mutual respect and trust**. The learning environment supports an interactive process. The teacher not only plans to learn from the students, but the students often assume the role of the educator.
4. The **use of local resources** assists in reducing costs without cutting quality. Nonformal education solves problems creatively through the use of existing resources.
5. Participation in nonformal education programs is seldom required. **Timeliness and usefulness** are critical concerns. Participants need information to solve problems and make decisions rather than to learn isolated skills or knowledge. Participants who cannot see immediate use or personal value are likely to leave the program.
6. Finally, many nonformal education programs exhibit a **lower level of structure**. To encourage and foster the learner-centered approach, the educator must not only be flexible but be willing to share control.

The Cooperative Extension Service prides itself in being a "grassroots" organization, taking its initiative from the needs and concerns of the people, sharing educational resources, and putting the newly acquired skills and knowledge to work for a better self and community. Extension's nonformal approach to education empowers the people to take charge of their lives.

THE NATURE OF ADULT EDUCATION

Facilitating the learning experience for adults necessitates an understanding of adulthood in conjunction with the learning process. Many adult educators support the idea that teaching adults is different from teaching children or adolescents. Adults bring to a learning situation different experiences and circumstances. An adult's ability to acquire new information may have more to do with life-style, social roles, and attitudes than with an innate ability to learn.

Adult education is emerging as its own discipline of study and as such many theories and ideas exist about how adults learn. Malcolm Knowles (1980, 1984) developed the most convincing model to show that teaching adults should differ from teaching adolescents or children. Knowles's "andragogical" or learner-centered model contrasts with "pedagogical" or teacher-centered models by arguing that adults differ from youth in a number of important ways that influence learning and consequently, how to approach learning. Knowles's (1984) model of andragogy is based on the following assumptions:

- adults tend to be self-directing (they determine their own learning needs)
- adults have previous experiences that can serve as a resource for learning
- adults' readiness to learn is frequently affected by their need to know or do something; therefore, they have a more problem-centered approach to learning as opposed to a subject-matter approach
- adult learners are seeking knowledge to fulfill societal expectations or roles (i.e., professional or family responsibilities). Their needs are for more immediate applicability than learning for future use.
- adult learners are more intrinsically motivated (the need/desire to learn is more important than grades, awards, etc.).

In 1988, Knowles modified his position regarding learning differences of adults and youth, stating that what he once believed to be unique characteristics of adult learners, he now believed are traits of all human beings that emerge as they mature. The one feature that does remain unique to adult learners is that adults by their very nature bring to the learning situation a wealth of knowledge. Despite the modifications, the andragogical model has significantly impacted the field of adult education. It still holds true, that most adult educators believe that teaching adults should differ from teaching children and adolescents.

As the discipline of adult education has emerged, others have both disputed and supported Knowles' early assumptions that teaching adults is different. Over time the following conclusions about adult learning needs have emerged. Since adults comprise a large percentage of Extension audiences, understanding the adult learner is important in developing sound programs for the Cooperative Extension Service. Cross (1983) argues that

- most adult learners choose educational offerings in which they can learn to do something that is of immediate, practical use to them. Traditional discipline-oriented subjects are likely to be popular only with degree-oriented learners.
- rewards such as better jobs and more pay motivate adults whose basic economic and educational attainments are low. Rewards related to personal fulfillment motivate only those persons whose basic necessities have been met.
- motivational factors may have a greater barrier to continued learning than previously thought. Environmental factors such as cost are frequently cited because they are socially acceptable. Adults seem to lack knowledge of actual barriers.
- adults respond more to their perception that the subject matter is credible and appropriate than to the location or schedule, regardless of the convenience factor.
- most adults respond more positively to interactive and active modes to learning than to passive modes such as listening or watching.

THE TEACHING-LEARNING INTERACTION

Education is a process comprised of both teaching and learning processes.

> *"Teaching and learning are really two separate functions. . . . If teaching and learning processes are to work effectively, a unique kind of relationship must exist between the two separate organisms—some kind of 'connection,' link, or bridge between the teacher and the learner"*
>
> (GORDON 1974, P. 3).

Although most agree that a relationship is present between teaching and learning, teaching can occur without learning and learning can occur outside the context of a specific structured teaching situation.

Leagans suggests that there are at least five factors that influence the teaching-learning interaction. Figure 6–1 illustrates the interrelationship between the five: the teacher, the learner, the subject matter, the physical facilities/environment, and the instructional materials and methods. Each factor can directly influence the quality of the learning experience.

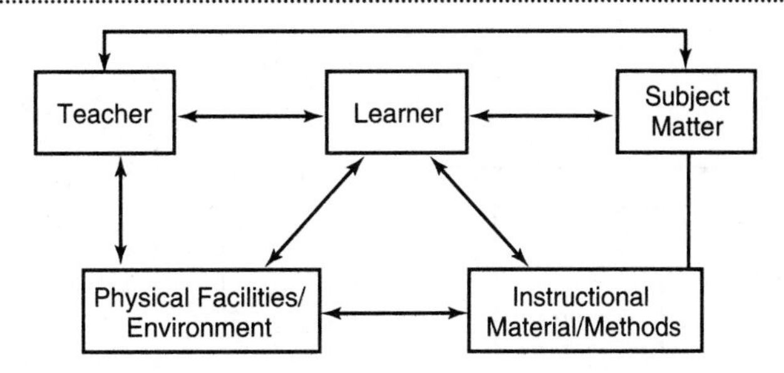

Figure 6–1 ***Major Elements in a Teaching-Learning Situation. Leagans, J. P.*** **Setting up Learning Situations.** ***Mimeograph. 4 pages. Cornell University, Ithaca, NY. Undated.***

Both the learner and the teacher will bring to the learning situation personal needs, interests, philosophies, and values; however, when these dimensions differ greatly the learning situation can be compromised. The subject matter must be relevant and appropriate for the intended audience. For example, a workshop on tailoring would not be appropriate for first-year 4-H members who are just learning to sew. The instructional materials and methods selected must also be appropriate to the learners' background and experience, relate to the stated program objectives for the subject to be addressed, and be able to be accomplished with the available facilities or learning environment. For example, information on financial planning could be taught anywhere a classroom environment could be established. However, if the intent were to utilize personal computers to maximize financial planning, then an environment where each learner could use or share a computer would be necessary. Also important to know would be which learners already had basic computer skills and access to computers on a regular basis.

Teaching as Communication

Thinking of teaching and learning as an interaction indicates that some exchange is occurring. Teachers and students are continually in the process of sending and receiving messages. Understanding the way communication occurs can help to understand some of the problems associated with teaching and learning.

The most basic communication model consists of three components: a sender (source of information), the message (content or subject), and the receiver. Frutchey (1966) devised a more elaborate model of the communication process to illustrate the Extension education model, see figure 6–2. A component addressed in this model is the selection of a channel by which the message is communicated. The channel identifies the specific methods used to convey the message. Extension uses many different channels including, among others, workshops, demonstrations, farm/ranch visits, discussion, and mass media. As the message is re-

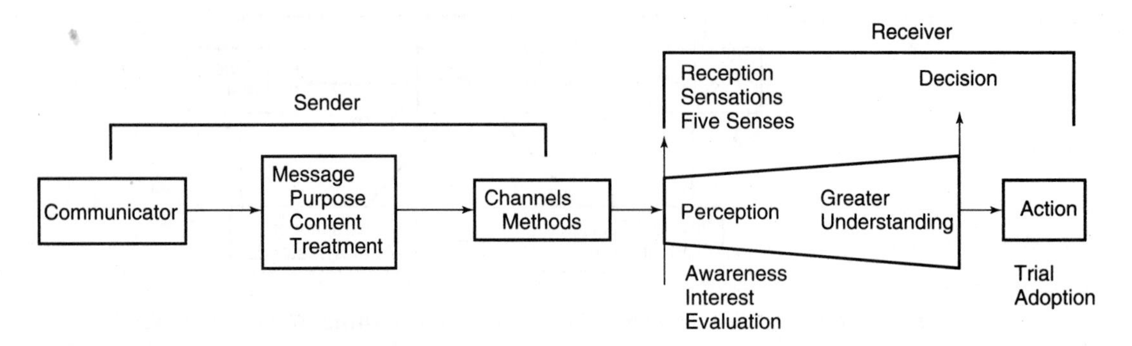

Figure 6–2 The Communication Process.

ceived, it is processed through one or more of the five senses (seeing, hearing, feeling, tasting, and smelling) to provide meaning and understanding. Once the message has been processed, it may be stored for future use or, if action is called for, a decision is made and action occurs. A primary function of Extension professionals is to share knowledge and to facilitate the adoption of new ideas and practices by clientele. This communication approach is called the diffusion and adoption of innovations.

The Diffusion Process. The adoption-diffusion model, originally developed to explain the educational processes that led farmers to accept new ideas, has been widely used in many areas. Rogers (1983, p. 34) defines diffusion as

> *the process by which an innovation is communicated through certain channels over time among the members of a social system. Diffusion is a special type of communication concerned with the spread of messages that are new ideas.*

An innovation is an idea or practice that is perceived to be new to the clientele group.

Adoption is the decision to accept or make use of the innovations as the best course of action. The adoption process consists of five steps:

- awareness
- interest
- evaluation
- trial
- adoption

Awareness of a new idea and interest in and evaluation of the topic occur when the message is first being received and processed. Trial and adoption are steps that may occur later as understanding and commitment increase. This can be illustrated through the following example. An Extension professional may use several communication channels (workshop, mass media, and personal contact) to create an awareness of an innovation—a new wheat variety that requires less water and has stronger pest resistance. Once the farmer becomes **aware** of the new wheat variety, enough **interest** may be generated to seek out additional information and knowledge. Armed with this new information, he begins to **evaluate** the risks and benefits of trying the new wheat variety. Ultimately a decision is made to try it or not. If the **trial** is felt to be successful, there is a strong chance that the idea will be **adopted** and over a period of time, change (increased profitability) will result. At each step of the process, it is the Extension professional's role to provide knowledge and resources to foster adoption of the idea.

One of the features of the adoption process is that it occurs over time. Frequently there will be a lengthy time lapse between the introduction of a new idea and its adoption on a widespread basis. Since the diffusion and adoption of innovations is a primary Extension function, success is often measured by the extent to which this time lapse is reduced or the rate of adoption is increased. Many factors influence the rate of adoption, including the characteristics of the innovation and types of people who may adopt it. Five characteristics of an innovation, as perceived by the potential adopters, greatly influence the rate of adoption. The five characteristics of an innovation are: relative advantage, compatibility, complexity, trialability, and observability. Rogers (1983, p. 15–16) defines these characteristics as follows.

Relative advantage: The degree to which an innovation is perceived as being better than the idea it supersedes in terms such as economic profitability, social prestige, physical convenience, and psychological satisfaction.
Example: Adoption of the new wheat variety may yield higher economic profitability due to higher yields from less pest damage and lower irrigation costs.

Compatibility: The degree to which an innovation is perceived as consistent with the existing sociocultural values and beliefs, previous experiences, and needs of potential adopters. An innovation that is not compatible will not be adopted.
Example: In some communities, education on topics such as parenting, birth control, and sexually transmitted diseases is felt to be the responsibility of the family or church and programs would not be attended if presented by other organizations or agencies.

Complexity: The degree to which an innovation is perceived to be difficult to understand and use. The more complex the innovation, the slower the rate of adoption.
Example: In many Extension offices the adoption of microcomputers was slow. Technology was perceived to be complex and in some areas support was not readily available.

Trialabilty: The degree to which an innovation may be experimented with on a limited basis. New ideas that can be tested on a trial basis have a greater adoption rate.
Example: Wheat farmers may be willing to try the new variety on one or two fields to assess it's merits but will probably be unwilling to completely convert to the new variety.

Observability: The degree to which the results of an innovation are visible to others. The easier it is for others to see the results of an innovation, the more likely they are to adopt it.
Example: If farmer A is successful in increasing yields and gaining larger profits with the new wheat variety, farmers B, C, and D will be more likely to plant the new variety as well.

All individuals in a community do not adopt an innovation at the same time. People can generally be divided into what are called "adopter categories" based on how quickly they adopt innovations. Usually a few will try the innovation initially, then a larger number will try it, and finally the remainder of the population will accept the new idea. The distribution of adopter categories can influence the rate of adoption of an innovation, see figure 6–3. The five adopter categories below are classified by degree of innovativeness, or how early an individual adopts new ideas.

Innovators: They are the risk takers and adventurers. They are eager to try new ideas and are the first to adopt an innovation. Innovators represent about 2.5 percent of the total population.

Early adopters: Early adopters are considered the opinion leaders and usually have earned great respect and position in the community. They represent the next 10 to 15 percent of the population to adopt an innovation.

Early majority members: Although they usually adopt new ideas just before the average member, they do so only after intense thought and deliberation. They represent about one third of the total population.

Late majority members: These people adopt new ideas shortly after the average. They tend to be skeptical in nature and adopt only out of economic necessity or social pressures. They represent about one third of the total population.

Laggards: These are the traditionalists of the group. They are rooted to the past and make decisions based on what was done previously. They tend to be suspicious of new ideas and innovators and change agents. Laggards comprise approximately 15 percent of the population.

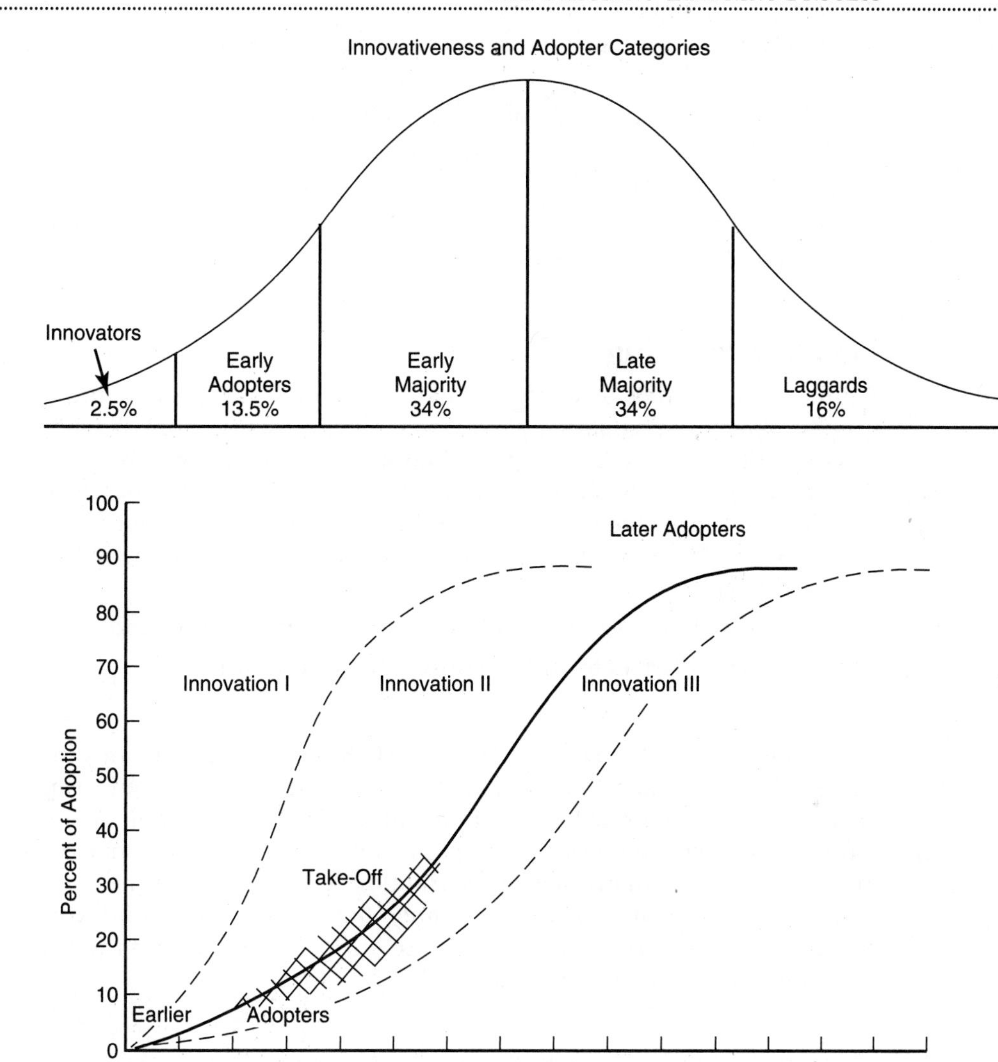

Figure 6–3   Innovativeness and Adopter Categories.

Each adopter category possesses unique characteristics and requires different strategies to influence. It is important that the Extension professional recognize individuals in each of these adopter categories to achieve successful adoption of innovations.

The diffusion process, built on the concept of change, embraces the Extension philosophy of helping to improve the quality of life by extending knowledge. As agents of change, Extension professionals assume responsibility for diffusing an innovation or idea and influencing its adoption. Havelock (1973) suggests that there are four roles an Extension professional can assume to influence adoption decisions. Those roles are:

- Catalyst—pressure the system to begin working on problems and issues
- Solution Giver—provide specific ideas for change
- Process Helper—assist in the processes of problem solving and decision making
- Resource Linker—bring together human, economic, intellectual, and political resources.

Effective Teaching

Thinking about one's own educational experiences can provide important insights about what makes a good teacher. Who was that one teacher that was the best teacher you ever had? Who was the worst? Why? What specifically did these individuals do or not do that determined their effectiveness?

There are three factors or predictors of teaching effectiveness:

1. Skill of the instructor in terms of presentation, knowledge, and clarity
2. Ability to organize and plan a course of instruction, and
3. Learner-instructor interactions. (Benton, 1982).

Knowledge of the Instructor. It goes without saying that an effective teacher needs to be competent in his or her specialized subject-matter area. However, technical competence alone is not sufficient. An understanding of teaching methods and philosophies assists the teacher in planning, delivering, and evaluating instruction. Understanding the audience, selecting appropriate teaching methods for the situation, and knowing how to effectively use those methods are critical. Knowing how to teach is as important as knowing what to teach.

The ability to plan instruction. Planning is a critical instructional skill. To plan a successful educational sequence, the instructor must recognize the need being addressed, be familiar with the many characteristics of the learners, develop instructional objectives that focus on the ends or purpose of the instruction, design instructional strategies to meet stated objectives, and evaluate to determine if the effort was successful. Designing a successful learning sequence considers: the content (what knowledge, attitude, or skill is needed to create the desired change?), the delivery (which instructional methods best address the topic?), and evaluation (were the stated objectives met?). A common temptation among educational planners is to write objectives based on what they (the instructors) plan to do and not on what the learning outcomes should be. Educational objectives need to clearly identify what the student is expected to know or do (performance) in specific and observable behaviors (measurement).

Educational programs should be designed with the end purpose in mind—specifically, the kind of change that is anticipated. Most programs fit into one or more of three categories known as Domains of Learning (Bloom, et al. 1956). The end goal of these three domains is increasing knowledge; changing attitudes, values, or beliefs; or developing skills. The three domains are described as **cognitive** (knowledge), **affective** (attitudes), and **psychomotor** (skills).

Each of the three domains of learning represent a hierarchial structure for learning. This approach describes a logical and sequential order of learning in which the preceding level must be met before the next level is addressed. The most recognized of the three domains is the cognitive (Bloom 1956) although the affective and psychomotor domains are also organized by levels. Bloom's taxonomy levels of cognitive learning and examples of learning at each of these levels are as follows:

Level 1–Knowledge:	learner memorizes information without understanding it
Level 2–Comprehension:	learner is able to identify relationships among ideas
Level 3–Application:	learner applies principles, rules, or procedures
Level 4–Analysis:	learner breaks communication down into parts and identifies relationships among the parts
Level 5–Synthesis:-	learner applies knowledge and skill to produce a unique and original product
Level 6–Evaluation:	learner judges the worth or value of the product based on specific criteria

Many different program planning models exist. Bennett's Hierarchy (Bennett 1975) is one model that has been used extensively in Cooperative Extension programming efforts. Principles of program planning, commonly used models, and directions for developing educational objectives were addressed in depth in Chapter 5, see figure 5–3.

Learner-Instructor Interactions. Equally important is the interaction that occurs between the learner and the instructor. Qualities that individuals find desirable in an effective teacher are enthusiasm, energy, clarity, honesty, variety in presentations, interest in the learner, approachability, use of examples and personal experiences, involvement of learners through discussion and activities, and fairness and constructive evaluation efforts. Rocheve and Blake (1986) shows how characteristics of excellent teachers fall into one of three categories:

Motivation Skills:

• Commitment	-believe in every student's ability to achieve
	-establish high expectations with realistic demands
	-are accessible outside of class
• Goal Orientation	-set personal goals
	-possess strong sense of direction
• Integrated Perception	-see "big picture" and integrate subject matter -link class content to real-life situations
• Reward Orientation	-have personal satisfaction in area -excite others -encourage student achievement

Cognitive Skills:

• Individualized Perception	-personalize instruction -recognize individuals within a group
• Teaching Strategies	-use a variety of approaches -exhibit enthusiasm -plan and organize well, encourage feedback
• Knowledge	-have expertise in content area to enable others to be effective
• Innovation	-remain current in subject area and translate it to instruction

Interpersonal Skills:

• Objectivity	-maintain discipline and respect in the learning environment
• Active Listening	-pay respectful attention to participants
• Rapport	-maintain approving and favorable relationship
• Empathy	-understand reality from another's perspective

THE LEARNING PROCESS

Directly related to teaching is the concept of learning. Learning relates to a change in knowledge, attitude, skills, or behavior. Learning has occurred when the learner has changed in some way. Learning is an active process that is not limited to an absorption of knowledge or information.

Many different factors influence why participants in the same educational situation learn differently. Factors influencing the learning process can be categorized into three areas: physical, emotional, and intellectual.

Physical factors concerning health and comfort can influence learning. Ill health or diminished sensory abilities can deter learning. For example, depending on the severity of the illness, the learner might experience a decrease in attention span and retention ability. Sensory factors such as impaired vision or hearing may limit the processing of information and knowledge. Physical discomfort can also influence rate and retention of learning. Sitting on an uncomfortable chair for an extended length of time will cause decreased interest and attention. Much Extension teaching uses the demonstration method. In a demonstration conducted outdoors, varying temperatures, wind, rain, etc. will influence learners' attention and absorption. Adequate lighting, good acoustics, proper ventilation, a comfortable temperature, room arrangement, and seating can all enhance or diminish the learning process.

Emotional factors are more complex. Motivation or desire to learn is a critical factor. When the learner is ready to learn, when the topic has meaning, relevance, or personal value, learning will take place more quickly. Readiness to learn is based in part on an assessment of the benefits as well as the disadvantages of not learning. In other words, the learner may ask, "how does this help me?" Readiness can also depend on the amount of previous learning. Degree of self-confidence can be linked to achievement. Learners who view themselves as capable of learning demonstrate a higher potential for learning. All learners bring to the learning situation certain values, beliefs, and attitudes. Learning involves changing attitudes, knowledge, or behavior, and change of any kind can be threatening. Thus, depending upon the emotional comfort and security of the learner, the learning situation may appear threatening and become less effective, or welcoming and become very effective.

Intellectual factors include previous experiences, passive versus active learning approaches, and learning preferences. Learners come from all backgrounds and age categories. Previous learning experiences can be both an asset and a liability. Previous learning can be an asset to learning by linking new knowledge and skills to ones already acquired. Drawing on experience and relating to existing knowledge helps to decrease anxiety and increase acceptance. Previous knowledge, however, can also be a liability. Some of the biggest barriers to learning are the notion of tradition, fear of change, and the need to "unlearn" habits or practices in order to learn new ones. Attitudes such as "we've done it this way for years," "it still works," and "if it's not broken don't fix it" can impede learning.

Learning is also enhanced when the educational process is active rather than passive. Learning is more permanent when the learner has taken part in solving the problem rather than simply given the solution. For example, when learning how to show a 4-H livestock project, develop a business plan, or prepare a low-fat meal, the person who learned by practicing it will be more likely to remember how to do it the next time, than the person who read about it or only observed the process. Repetition and reinforcement also enhance learning experiences. In early Cooperative Extension history, Seaman Knapp utilized the demonstration method extensively and with great success. The 4-H motto is "Learn by Doing." Over time, these approaches have proven effective and should be used as often as is feasible.

There is much information about what are sometimes called learning styles—the preferences individuals have for how they learn. Learning style refers to the characteristics and preferred methods of gathering, organizing, and interpreting information. All learners have preferences about their involvement in a learning situation. There are certain circumstances that, when present, minimize anxiety and increase learning for some individuals but not for others. Preferences may be noted regarding a variety of factors. For example:

Do you prefer absolute silence or do you enjoy music or other background sounds?

Do you like to work in a bright environment or one that is soft and dim?

At what time of day are you at your peak?

Do you prefer to work in groups or alone?

Given the option would you prefer reading a pamphlet, listening to a lecture, watching a demonstration, or conducting an experiment?

No one single style has been found to be better than any other, nor does any one style lead to better learning. And it is likely that in any learning situation, the audience will be comprised of learners who possess a variety of different learning styles. It is nearly impossible for any instructor to match his or her teaching to all the possible learning styles the students possess; therefore, a variety of instructional and evaluation strategies should be used to address the diversity of the group.

WHY PEOPLE PARTICIPATE IN EDUCATION

One of the basics in teaching adults and youth is understanding their motivation or reason for participation. Maslow's "Hierarchy of Needs" (1970) has become a classic model of human behavior and is widely used in a variety of contexts. Maslow's theory states that people are motivated to satisfy specific physiological needs. These needs range from the very basic and physical to very complex and psychological. Maslow's hierarchy consists of five levels:

Level 1–Physiological:	basic physical needs for food, water, air
Level 2–Safety:	the need to be safe from harm, to have security
Level 3–Social:	the need for affiliation or closeness with others, to be liked
Level 4–Esteem:	the need to be recognized as a person of value, to be rewarded
Level 5–Self-Actualization:	the highest need; a person will not be ultimately happy unless he is doing what he is fitted for

Maslow noted two observations in his study of human behavior. The first is that humans are basically "wanting creatures," so as soon as they have satisfied one level of need, they move on to the next. Likewise, if a basic need is not met, all other needs become unimportant and we regress on the hierarchy until it is met. His second observation was that a need that has been met is no longer a motivator.

Extension programs relate to all levels of Maslow's hierarchy. A few examples for each of the levels include:

- **Physiological:** knowledge and survival tips for disaster victims (flood, tornado), water quality
- **Safety:** farm safety programs, proper disposal of household hazardous materials
- **Social:** participation in programs and activities such as 4-H clubs
- **Esteem:** the many and diverse opportunities to volunteer time and talents to Extension programs.

Maslow's theory provides a general guide to human behavior and participation. Houle (1961) describes motivations of three types of adult learners:

Goal oriented: individuals who use education as a means of accomplishing well-defined objectives (specific problem to be solved)

Activity oriented: individuals who take part because they want to be involved in the activity or process and not necessarily because they have a strong interest in the subject or content (for fun and/or to fulfill social needs)

Learning oriented: individuals who seek knowledge for its own sake (strong interest but not necessarily a need).

Understanding motivations for participation can assist the Extension educator in planning and implementing programs that closely reflect the interests and needs of participants. In addition, understanding barriers to participation is helpful for planning strategies to overcome them.

Barriers to participation in adult education programs can be categorized as situational, institutional, or informational. Most barriers fall into the **situational** category—those factors that relate to a particular time in relation to the social and physical environment. Examples include busy schedules, lack of child care or transportation or money. **Institutional** barriers are those that originate from the institution itself, for example, inconvenient scheduling or restrictive locations. **Informational** barriers may be due to lack of information about the organization or program. Many people in the United States are not only unfamiliar with the Cooperative Extension Service, they are totally unaware of its existence or programming.

PLANNING FOR SUCCESS

The teaching-learning interaction is a complex process with several factors that need to be considered for a successful effort.

- **Know the audience.** Extension audiences, in particular, are very diverse. The successful educator will attempt to learn as much about the audience as possible. The more information available, the greater the chance of creating a learning experience that is meaningful and appropriate. For example, when working with a limited-resource audience, recommendations and procedures can be adapted to fit within budget restrictions.
- **Plan, Plan, Plan.** Be prepared. The effective teacher never approaches the learning situation unprepared. He or she establishes instructional objectives based on an identified need and chooses teaching strategies that are appropriate for the level of the audience, the nature of the subject, and the resources (time, expertise, money, and facilities) available.
- **Establish a Positive Environment for Learning.** The learning environment consists of both physical and psychological factors. Optimum learning occurs when all conditions are considered.
    1. Physical factors include:
       -good lighting and visibility
       -comfortable seating
       -room arrangement suitable for learning activities
       -good acoustics
       -accessible to the disabled
       -comfortable temperature—not too hot or cold
    2. Psychological factors involve establishing an environment that is comfortable, based on mutual respect and trust, and encourages interaction. This can be accomplished by:
       -becoming acquainted with students, greeting them, and if possible learning their names
       -involving students in their own learning
       -providing reinforcement and encouragement, avoiding put-downs
       -recognizing differences and respecting the uniqueness of each student
       -being accessible

SUMMARY

Teaching and learning is an interactive process that occurs in many different settings and situations. Factors such as teacher effectiveness, barriers to learning, physical environment, motivation of the learner, and subject content are a few of the circumstances that can enhance or diminish the learning situation. The Cooperative Extension Service is the world's largest adult education nonformal program. One of the primary functions of Extension is the diffusion and adoption of innovations. The diffusion process in a unique communication method for sharing knowledge about new ideas. The successful Extension educator carefully considers the many influences on the teaching-learning environment when planning and conducting Extension programs.

DISCUSSION QUESTIONS

1. What are some innovations that have been diffused in agriculture in the last ten years? What has been the role of Extension in this process?
2. How has agricultural education changed over the last decade? What influence has this had on the education system?
3. Who was your best teacher? Your worst? Why?
4. What responsibilities does the learner have in the learning environment? The teacher? What rights?
5. What are some strategies for eliminating barriers to participation?

REFERENCES

Bennett, C. 1975. Up the Hierarchy. *Journal of Extension.* March–April, Vol. 13, 7–12.

Benton, S. E. 1982. Rating college teaching: criterion validity studies of student evaluation-of-instructor instruments. AAHE-ERIC/Higher Education Research Report Number One. Washington, D.C.: American Association of Higher Education.

Bloom, B. ed. 1956. *Taxonomy of educational objectives.* New York, NY: David McKay.

Bloom, B. S., Englehart, M. P., Fuest, E. J., Hill, W. H., and Kratwhol, P.R. 1956. *Taxonomy of Evaluation Objectives Handbook 1; Cognitive Domain.* New York, NY: David McKay Co.

Boone, E. J. 1985. *Developing programs in adult education.* Englewood Cliffs: NJ: Prentice-Hall.

Coombs, P. 1973. *New paths to learning for rural children and youth.* New York, NY: International Council for Educational Development.

Cross, P. K. 1983. *Adults as learners.* San Francisco, CA: Jossey-Bass.

Darkenwald, G. G., and Merriam, S. B. 1982. *Adult education: foundations of practice.* New York, NY: Harper Collins.

Etling, A. 1975. *Characteristics of facilitators: the Ecuador project and beyond.* Amherst, MA: Center for International Education, University of Massachusetts.

Frutchey, F. The Teaching-Learning Process. 1966. In Sanders, H. C., *The Cooperative Extension Service.* Englewood Cliffs, NJ: Prentice-Hall.

Gordon, T. 1974. *T.E.T.: teacher effectiveness training.* New York, NY: Peter H. Wyden.

Havelock, R. G. 1973. *The change agent's guide to innovation in education.* Englewood Cliffs, NJ: Educational Technology Publications.

Houle, C. O. *The inquiring mind.* 1961. Madison, WI: University of Wisconsin.

Khan, J. 1989. *Design of a workshop to train P.S.U. faculty as international consultants in youth development.* Professional paper in Extension Education. University Park, PA: Department of Agricultural and Extension Education. The Pennsylvania State University.

Knowles, M. 1980. *The modern practice of adult education: from pedagogy to andragogy.* Revised and updated. Chicago, IL: Follett.

Knowles, M. 1984. *Andragogy in action: applying modern principles of adult learning.* San Francisco, CA: Jossey-Bass.

Leagans, J. P. Undated. *Setting up learning situations.* Mimeograph. Ithaca, NY: Cornell University. (As published in Prawl, W.; Medlin, R.; and Gross, J. 1984. *Adult and continuing education through the Cooperative Extension Service.* Columbia, MO: Extension Division, University of Missouri.

Maslow, A. H. 1970. *Motivation and personality.* 2nd. ed. New York, NY: Harper and Row.

Rasmusen, W. D. 1989. *Taking the university to the people.* Ames, IA: Iowa State University Press.

Rogers, E. M. 1983. *Diffusion of innovations.* 3rd. ed. New York, NY: The Free Press.

Roveche, J., and Blake, G. 1986. Profiling excellence in America's schools. *Current Issues in Higher Education*:102–107.

Websters new world dictionary of the American language. 1994. New York, NY: World Publishing.

CHAPTER

"What a man hears, he may doubt. What he sees he may possibly doubt. But what he does himself he cannot doubt."
—*Seaman Knapp*

Extension Teaching Methods

Key Concepts

- The Extension education process
- Classification of Extension teaching methods
- Individual, group, and mass-media methods explained

Useful and Practical Instruction

Two words have characterized the educational work of Extension from its beginning. The words *useful* and *practical* have been emphasized since Seaman Knapp selected the first agents from among the most "practical-minded" citizens of the county. While Extension teaching methods have changed during the near century of its formal existence, providing instruction has remained the major function and primary responsibility of Cooperative Extension work.

In the early days it was necessary to give highly individualized instruction, which tended to be of a service nature. As Extension developed, its methods evolved through various stages of individual, family, group, and community emphasis. Today, the Extension professional assists with planning and organizing community resources as well as teaching to address local issues. "Service" is still achieved, largely as a result of teaching. Educational programs are designed to influence people to make desired changes in their behavior through an adoption process that will contribute to a better life. Changes can be taught by the Extension professionals in three ways:

1. By increasing the amount of useful information or understanding available
2. By introducing and demonstrating new or improved skills, abilities, and habits
3. By fostering more desirable attitudes and aspirations

Thus, an effective educational program contributes to the individual's understanding, helps to improve his or her abilities, and helps develop more desirable attitudes. Extension education implies the need for adoption of new practices, skills, and techniques and emphasizes the need for change as a result of learning.

Extension professionals are known collectively as change agents who work to help individuals apply new technology, methods, or practices. To accomplish this effectively, these change agents need both breadth and depth of training. A knowledge of the physical and biological sciences and an application of the behavioral sciences is essential. Extension educators work through people to encourage the application of knowledge to the problems of its clientele. The application of the behavioral sciences is used to know and understand why people behave as they do so that effective learning experiences can be designed. In many cases, one method of instruction is not enough to assure adoption and hence the educator must use complementary and reinforcing methods of instruction and multiple communication systems (Waldron and Moore 1991). These efforts to influence change through instruction are known as the Extension education process, which is explained in greater detail in Chapters 5 and 6. The key elements of this process are summarized in figure 7–1.

Before any teaching takes place there must be a sound **program** based upon the needs of the people. This is developed based upon a situational analysis of the people's needs and interests, educational levels, social customs, and the economic and social conditions of the community.

Second, there must be a **plan of work** or plan of action that puts the educational program in operation. This involves setting specific objectives stated in terms of the behavioral change desired, selecting teaching methods, scheduling activities, and dividing responsibilities among Extension professionals for each program area.

Third, the plan is **implemented**. The carrying out of the teaching methods and related activities in the plan of work requires a systematic, patient, and persistent effort on the part of Extension educators involved.

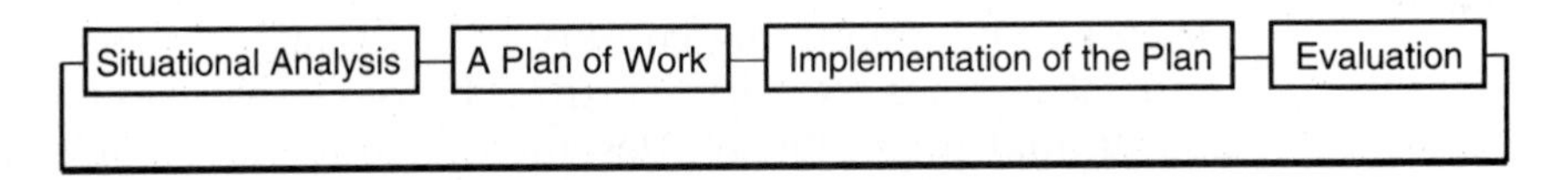

Figure 7–1 The Extension education process involves distinct and interrelated stages

Finally, progress and accomplishments are **evaluated**. Measurements to determine if the objectives have been reached make it possible to adjust the methods and activities to developing situations. Evaluation helps in the revision of the program at intervals to keep ahead of problems. Evaluation is the basis to improvement of the conduct of Extension work. If the evaluation results reveal that there is need for further work, then the process may begin again. Experienced professionals plan, learn, and evaluate continuously throughout all the stages of this process. A further explanation of evaluation is found in Chapter 8.

EXTENSION METHODS

The choice of teaching methods and activities will have a direct bearing upon the success of the entire educational process. Success or progress at each stage is dependent upon the appropriateness of the methods and procedures used in carrying out the preceding stage. The ability to choose and use those methods best adapted to particular objectives is a measure of an Extension educator's effectiveness. During almost a century of Extension work, a variety of ways to influence people to change their practices have been developed. Each way has its advantages and limitations, and each is especially adapted to a particular situation. Knowing when, where, and how to provide each learning experience to the best advantage is the mark of an Extension professional.

Methods Classified by Nature of Contact

Teaching methods used by Extension personnel were classified by early educators according to the number and nature of the contact (Wilson and Gallup 1955), see figure 7-2. One such classification is **individual contact**—a farm or home visit by an Extension educator or a client's visit to the office or telephone call requesting specific assistance. Individual methods are those in which the educator provides personal consultation to the client.

INDIVIDUAL CONTACT	GROUP CONTACT	MASS CONTACT
Farm or Home Visit Office Visit Telephone Calls Personal Correspondence Result Demonstrations	Meetings Method Demonstrations Leader Training Tours and Field Days Organized Clubs Camps Community Forums Short Courses Workshops Teleconferencing	News Stories Radio Television Appearances Newsletters Publications Interactive Conferences Computer-Aided Instructional Learning Satellite Programs Exhibits Telephone Message Answering Systems Internet

Figure 7-2 Classification of teaching methods by nature of contact

A second category involves teaching a number of people assembled in a **group** or in a number of groups. This classification includes all kinds of meetings held for many purposes such as training meetings, demonstration meetings, tours, and the like.

The third category involves the various **media** used by Extension professionals to disseminate information and influence large numbers of people. This includes the use of radio, newspapers, television, exhibits, and electronic technology.

Methods Classified by Form of Communication

Teaching methods may also be classified by the form of communication used, see figure 7–3. Bulletins, leaflets, news articles, personal and circular letters all depend largely upon the **written** word. The methods that depend upon the use of the **spoken** word characterize the variety of special and general meetings held by Extension educators. Farm or home visits, office visits and telephone calls also involve oral communication, while the radio is completely limited to the oral method of presentation. **Visual** methods of teaching, which depend almost entirely upon eye appeal, include result demonstrations, exhibits, posters, slides, charts, and similar aids. Visual aids are frequently used to supplement the spoken and written word to build attention, maintain interest, or increase teaching effectiveness. Method demonstrations, meetings, and television programs combine visual material and oral presentation.

Methods Classified by Function

The use of technology is becoming more important to Extension educators and the Extension system. An Extension Committee on Organization and Policy (ECOP) task force on technology classified Extension work by these three functions. The first function is **information delivery** where information flows through many channels of communications from the Extension Service to its clientele. This may include news articles, meetings, or personal consultations. The second function is **educational program delivery**. Educational programs are prepared and delivered by Extension specialists and agents to upgrade the knowledge, skills, and capabilities of the clientele. A program may involve many different activi-

WRITTEN	SPOKEN	VISUAL
Bulletins	Meetings	Result Demonstrations
Leaflets	Farm/Home Visit	Exhibits
Fact Sheets	Office Call	Posters
News Articles	Telephone Call	Slides
Personal Letters	Radio	Videos
Newsletters		Charts
		Television

Figure 7–3 Classification of teaching methods by form of communication

ties or learning experiences. Together these learning experiences become a program focused upon special audiences, special needs, and special problems. The third function is **problem solving**. Clients turn to the Extension Service for expertise, knowledge, and skills needed to solve individual and group problems arising on their farms or ranches and in homes and families. These three functions may overlap, depending upon the type of program and the approach taken by specialists and agents. The relationship of media technologies to these three functions depends upon the audience size and type.

Waldron and Moore (1991) classify Extension education by the techniques employed. They define *techniques* as the way in which the Extension educator establishes a relationship between the learner and the learning task. They categorize the techniques used by Extension professionals as

1. **Information-giving techniques,** which include lectures, question-and-answer sessions, panel presentations, and debates
2. **Skill-acquiring techniques,** which include workshops, laboratories, demonstrations, role-playing, drill, case study, and simulations
3. **Knowledge-applying techniques**, which include group discussions, workshops and laboratories, various forms of group activities, and case studies

It is apparent that a variety of techniques, functions, and forms of communication are used by the Extension educator. An expanded explanation of the types of contact Extension educators have with clients will help the novice better understand how and when to use each of these methods. Classification by the nature of contact will be used to categorize these methods as described by Sanders (1966), Kelsey and Hearne (1955), Wilson and Gallup (1955), and Hussey (1985). While not every method used by Extension professionals can be included, the following provides an explanation of the more commonly used methods.

INDIVIDUAL CONTACT TEACHING METHODS

While the individual contact can limit the total number of clients reached, it is one of the most effective methods of teaching. Some of the more commonly used individual methods are the farm or home visit, the office call, telephone calls, personal correspondence, and the result demonstration. One-on-one instruction supports one of the essential principles of motivation and learning. Adults are motivated to learn when they ask for information and assistance on a particular problem. These individuals have a desire to learn and are easy to teach.

Personal contact is vital to Extension work. Extension educators often make personal visits to clients to assess a situation, diagnose problems, and recommend solutions. Solutions are normally provided by the agent at the time of the request for assistance; however, a cadre of specialists is available for consultation on more complicated problems. Electronic technology allows Extension professionals to store and retrieve information and access specialists' input in a highly efficient manner. The effect is to amplify the expertise of agents in problem-solving activities.

Farm or Home Visit

The visit of the Extension educator to a farm or home serves many purposes but generally it is to give or obtain information. It may also be for the purpose of securing a cooperator or demonstrator, arranging a meeting, or discussing a local club activity. The personal visit may be to create good public relations with officers of local organizations, elected officials, or other key individuals. The farm or home visit may be merely to extend the agent's acquaintance or it may be part of a planned effort to interest those who do not participate in organized Extension teaching activities and are not reached though the mass media.

The farm or home visit provides an opportunity to work out practical solutions to specific problems. The visit makes it possible to modify general research recommendations in order to fit specific situations. It also provides an opportunity to arouse interest in improvements not yet recognized by the individual as desirable.

If the visit is for the purpose of obtaining information, that information can be interpreted more effectively because of the agent's first-hand knowledge of the circumstances involved. Personal knowledge of the farm or home conditions can be obtained only through contacts at the farm or in the home. These visits are essential to program determination and the selection of effective local leaders. There is a danger, however, in concentrating visits on the most progressive families and neglecting those whose personal contact is most needed. Farm or home visits contribute greatly to the effectiveness of the teaching process carried out through meetings, the media, or newsletters.

Office Visits

The office visit is a method of direct personal contact between the Extension educator and the individual desiring information or assistance at the Extension office. Typical callers are seeking an answer to a current problem, technical advice or information, or assistance with the organizational management of a local club. Because the visitor recognizes there is a problem to be solved and has a strong desire to solve it, the "climate of readiness" is more favorable for learning than with many other methods of contact. Confidence in the Extension educator as a reliable source of information is not to be taken for granted. The atmosphere of the office and the manner in which the desired information or advice is given has an important bearing on repeat calls. The office secretary plays a very important role in assisting the client and ensuring a satisfactory experience. It is very important that the office visitor be assisted in a friendly, courteous, yet business-like manner.

Telephone Calls

The telephone is another important means of person-to-person communication linking the agent to the community. The telephone is used to request specific subject-matter information and to facilitate other teaching activities. By the telephone, appointments are made, meetings scheduled, programs arranged, bulletins requested, progress checked, and a host of other business transacted for the maintenance of programs. Some counties also use the phone for prerecorded messages or public service announcements. Local publics can call for the latest consumer tip or select from a menu of information options. Telephone requests for information are often answered by mailing bulletins or other materials, but some problems may also prompt a visit by the agent to examine a situation.

Much time can be spent on the telephone with the state Extension office discussing administrative matters with the district supervisor and answers to unusual problems with the subject-matter specialist.

Personal Correspondence

The telephone request has almost replaced personal correspondence but the mail is still one method of individual contact. In many cases, requests by correspondence are from individuals who have a generic request such as "anything on growing roses." Such correspondence is generally answered with a mailing of bulletins or leaflets on the subject of interest.

County agents still receive a large volume of correspondence each day from the state Extension office on a variety of subjects. Much of this relates to monthly or annual reports or routine specialists' publications or other resource materials.

Result Demonstration

A result demonstration is a demonstration conducted by an individual under direct supervision of an Extension educator to prove the advantages of a recommended practice or combination of practices. It involves careful planning, a substantial period of time, adequate records, and comparisons of results. It is designed to teach others in addition to the person who conducts the demonstration. Result demonstrations establish proof that the improved practice advocated is applicable locally. Result demonstrations do not discover new truths, but they show to what extent the research findings of the state experiment stations, the United States Department of Agriculture, and other agencies apply to local conditions. Variations of the result demonstration are the on-farm research project, applied research, or verification trials.

On-farm demonstrations have served as one of the most effective Extension education tools ever developed (Hancock 1992). The need for such demonstrations was first recognized by Seaman Knapp when he found that farmers would not change their methods of farming unless they conducted the demonstration themselves on their own farms under ordinary conditions.

While result demonstrations are not limited to agriculture, it is the program area that uses this method of teaching the most frequently. On-farm demonstrations are used as a problem-oriented approach to agricultural research that begins by diagnosing the conditions, the practices, and problems of a particular group of farm producers. To be effective, result demonstrations must be carried on systematically to prove that the recommended practice is definitely superior to the one it is to replace. Accurate records are essential—records of labor, materials, costs, and results.

A result demonstration may deal with a single practice, such as the use of certain fertilizers for a growing crop; or it may include a series of practices, such as those involved in the management of a total farm enterprise. Some demonstrations are completed in a short time while others continue a year or more. Simple, clear-cut demonstrations are preferable for use in teaching.

The chief purpose of the result demonstration is to establish confidence in the Extension educator and the demonstrator. The fact that the practice has been successfully carried out by a local citizen builds confidence in the soundness of the practice and the Extension agent who directed the demonstration. The result demonstration is the most expensive method of Extension teaching due to the large amount of time and travel required of the agent to plan, start, and monitor the demonstration. If confidence in a community is already adequate, the agent may likely use a leading producer in the community who is already using the improved practices as an example or illustration. These leaders can become much like demonstrators because they are already carrying out the recommended practice. They can also be used effectively by the agent for teaching purposes.

Demonstrators should be selected from communities in need of the recommended practice. Persons who are selected to be demonstrators should be dependable, honest, unbiased, and have the confidence of the community. They must be public-spirited and willing to take the time and give the attention needed to take the demonstration to successful completion. The demonstrator should be a good manager, successful in personal affairs and finances, and be progressive, one who is ready to try new methods or practices. It is desirable that the demonstrator be energetic, active, and interested in community affairs. If this is an agricultural demonstration, it is suggested that the demonstration plot be easy to reach and located along a highway or main road.

Demonstrations designed for low-income families should be conducted on farms or in the homes of low-income families where the applicability can be demonstrated and the cost of adopting the practice is minimal. The selection of the demonstrator is especially important because the community will look to its most progressive families to help solve its problems. These demonstrators can be very influential in persuading local low-income citizens to adopt the recommended practice. The same principle applies when working with other groups, such as part-time farmers, single parents, or the elderly.

Convincing people too often leads to good intentions, but no action. People must be stirred to action. This means that appeals have to be made to basic human needs. Seeing what their neighbor has achieved, how it was achieved, and how they can have similar benefits is one of the strongest forces for motivating people to make desirable changes.

Change can be quickened by demonstrating how profitable or desirable a plan or remedy is. It may be necessary to overcome objections by proving the demonstrated remedy is easy to understand and adopt, convenient, low in cost and satisfying. Persuasive demonstrations such as weight loss programs or pasture improvement can have visible results. Another strong motivating influence for change is the desire to avoid losses, such as from plant and animal diseases or insect pests. Result demonstrations can illustrate ways of avoiding such losses and may serve as the catalyst for community action. They may also make families aware of potential and unrecognized losses. People are likely to repeat the experiences or practices that satisfy them and drop the practices that fail to satisfy them. The results, in relation to cost and effort, must be satisfactory if clients are to continue to use a practice.

While individual instruction is very effective, it also has some significant disadvantages. These points are summarized in figure 7-4.

ADVANTAGES	DISADVANTAGES
Provides first-hand knowledge of local problems	Cost per contact is higher than other methods
Establishes a climate of readiness for learning	Limits the number of total contacts that can be made by local Extension professionals
Builds confidence in the agent as a reliable source of information	Neglects some clientele who need assistance if caution is not taken to visit with representative families
Contributes to the selection of local leaders, demonstrators, and cooperators	Requires good planning to offer timely instruction for the farm/home visit
Aids in contacting individuals not normally reached in Extension activities	Removes the educator from the actual situation when individual contacts are limited to office or telephone information.
Is an effective teaching method. Individual visits are normally a quick and easy way to disseminate information.	Presents an opportunity for a communication problem when the question is not understood or the answer provided is misunderstood
Develops good public relations	Projects a poor public image when correspondence or phone calls are not answered promptly
Provides immediate feedback to questions and/or problems	Requires many hours of planning and follow-up if a result demonstration is to be successful
Provides local proof about research recommendations	Requires good time-management skills to handle constant requests for assistance

Figure 7–4 Advantages and disadvantages of individual teaching methods

GROUP METHODS

With improved transportation and technology, educational efforts in Extension work have shifted to group methods. Group methods are an economical way to reach larger audience numbers. Usage of each particular group method varies from state to state as does the terminology associated with the methods.

The Method Demonstration

The method demonstration is an especially effective way of teaching how to do something. Demonstrating, along with judging and exhibiting, became a vital part of early Extension education activities in youth work. Demonstration contests, individually and with teams, were looked upon as stage performances of practical teaching in agriculture and home economics. The method demonstration focused on what constituted good demonstrating so that all club members could acquire the skill. A club member was not only to know how to do something well but was to know how to teach well enough that others could learn the skill.

As a current teaching method, the method demonstration is a how-to-do illustration given by the Extension educator or trained leader to teach a skill to a group. The demonstration is a step-by-step explanation of a procedure. The demonstration involves diagrams, charts, and illustrative materials combined with the oral explanation. The learners watch the process, listen to the explanation, and ask questions during or at the close of the demonstration to clear up points of uncertainty. The combination of "seeing" and "hearing" makes a strong impression, further strengthened by practice through participation in the demonstration. Certain subjects lend themselves to demonstration better than others. Practices and skills, such as "how to" iron a shirt, are taught well by demonstration. More complicated practices like making a dress may involve many method demonstrations since dressmaking involves a series of "how-to" steps that culminate in a final product. The method demonstration should not be confused with the result demonstration. A result demonstration planned to compare fruit production in a double-row system (plastic mulch, drip irrigation) with a standard single-row system (no irrigation) may include a method demonstration on how to prune the fruit trees, see figure 7–5.

Use of local leaders to replicate method demonstrations, after appropriate training by the Extension educator, is a method used in the organized club programs of 4-H and Family Community Education (FCE) Clubs. This repetition of the method demonstration by trained leaders increases the impact of teaching without significantly increasing Extension costs.

General Meetings

General meetings include all kinds of meetings, other than method-demonstration and leader-training meetings, held by Extension agents. A

RESULT DEMONSTRATION		METHOD DEMONSTRATION	
Conducted by:	Farmer, homemaker, club member	**Conducted by:**	Extension employee, 4-H or FCE project leader or club member
Designed to teach:	Others in addition to persons conducting the demonstration	**Designed to teach:**	Persons present
Where:	On the farm, in home or geographic area	**Where:**	At training meetings, general meetings, or on TV
Duration:	Usually several weeks or months	**Duration:**	Length of meeting only
Purpose:	To establish local visible proof of the advantages of one practice over another or of a new practice	**Purpose:**	To teach a skill or method. To show step-by-step how to carry out a practice

Figure 7-5 Comparison of method demonstration and result demonstration teaching methods

great variety of general meetings are used by Extension, some of which are targeted to specific audiences while others are targeted to teach certain subject matter. The size of the meeting may range from only a small number of individuals (such as a committee) to hundreds of individuals at a community meeting. The focus of a meeting may be a neighborhood, community, county, or state. The method of presentation may be a formal talk, an informal discussion, the showing of a video, or a combination. The location may be a home, a community meeting room, or a field. Local use and custom often determines whether the meeting is called a workshop, special interest meeting, school, institute, forum, conference, clinic, discussion group, club meeting, project meeting, or planning meeting. Some meetings take the name of the meeting objective, e.g., program-planning meeting, annual meeting, achievement day, rally, tour, conference, or short course.

Meetings at which method demonstrations are given make it possible for groups of people to learn new or improved skills. General meetings make it possible for large numbers of people to acquire subject-matter information. The group approach makes it possible for participants to share knowledge and experience with others, thereby strengthening learning. General meetings may also involve objectives only indirectly connected with the dissemination of information on better practices. Such objectives include the development of local leaders, understanding of agricultural policy or other public issues, and recreation and social contacts. Advisory meetings for the purpose of determining the Extension programs for a club, community, or county are indirectly associated with teaching and learning. Such meetings serve to attract attention, arouse interest, and strengthen the conviction that something should and can be done about problems of vital concern to individuals.

Short Course. The lecture presentation is used extensively by Extension educators and subject-matter specialists to present authoritative or technical information. This method of teaching is often used at short courses. The short course generally has four to six lessons of instruction on a specific subject. The speaker presents specific subject-matter information to a particular audience, with communication flowing mostly from speaker to audience. At times, questions at the end of the lecture establish some interaction between the speaker and members of the audience. Some short courses use a combination of lecture and demonstrations. During a short course, the emphasis is focused upon in-depth instruction on a particular subject. An example is a landscaping short course in which all the instruction is related to establishing and maintaining lawns and ornamental plants.

Special Interest Meeting. This method is very similar to the short course except it may be only one or two meetings in duration. While the short course emphasizes the subject matter being taught, the special interest meeting focuses on the audience. Special interest meetings are held for audiences who would have "special interest" in the subjects being taught. An example would include new child care laws for an audience of child care providers.

Workshop. The workshop is a series of meetings for the intensive study of, work on, or discussion of a specific topic. A workshop usually includes demonstrations or lectures by an instructor followed by hands-on participation or practice by the workshop participants. A tailoring workshop would be one example. The instructor would discuss and illustrate how to finish a bound button hole after which the participants would practice the skill.

Tours and Field Day Meetings. The tour is a series of field and demonstration meetings arranged in sequence. The scheduling of several field day meetings gets wide publicity and attracts more than usual attention and interest. A tour may be devoted to a single project or activity or the cumulative effect of several demonstrations. Another type of tour may include a variety of subjects selected to attract wide participation and acquaint the public with important aspects of the Extension program. The practical setting of the field meeting, usually with small groups, stimulates informal discussion about the applicability of a new practice. At scheduled stops, the Extension agent can illustrate a practice and help the participants inspect progress or witness the outcome of a result demonstration. Tours may also be used to create an awareness of certain practices or new technologies. An example would be a tour established to illustrate new varieties of crops tolerant to insects or diseases.

Camps. Camps have been a popular method of group instruction for many years. Used mostly in youth work, the camp setting may include rustic cabins, modest buildings, or hotel-quality lodging quarters. The camp program is a combination of instructional activities with planned educational and recreational activities outdoors. The camp program's length varies to fit the focus or curriculum to be taught. The focus of the instruction may be topic specific, such as environmental stewardship, or it may be a series of activities designed to reach an objective. The outdoor setting can also be used to teach leadership and decision making. The 4-H Outdoor Leadership Camp, Forestry and Wildlife Camp, and Space Camp are examples of such camping experiences.

Clinics. A clinic is a learning experience through examination and treatment. A clinic usually has an "expert" or specialist who assists in identifying and analyzing a problem. The clinic may be a short, intensive session of group instruction on a specific skill or field of knowledge. The purpose is to advise about a specific problem. An example would be a cholesterol screening clinic. After readings were discussed, advice would be offered about food choices or referrals to medical doctors might be made.

Group Forum. Discussion at a group forum is a process whereby two or more persons express, clarify, and share their knowledge, opinions, and feelings. The combined and cooperative thinking of several persons is likely to be superior to that of one individual. Group discussion is used to present new technical information. It is employed to develop desirable attitudes, relate knowledge to experience, deepen understanding, reach agreement, and plan action. The role of the discussion leader is to stimulate participation by all the members of the group. The discussion may follow a lecture presentation, video presentation, symposium, panel discussion, interview, or community forum or debate. Specialists in the community development program area use this approach often to solve community issues. The forum may be held with community leaders, councils, or advisory groups as the participants. An example of a community forum would be a gathering of local political leaders, civic planners, and university researchers to discuss a proposed county landfill.

4-H Activities. Educational activities play a vital role in helping to maintain interest in 4-H or FCE work and in instructing the membership. County, state, and national activities are conducted each year to maintain interest in project areas and the program in general. These activities may be educational in nature, or competitive, or a combination of the two. Examples of competitive activities include county activity days with competitive talks and demonstrations in public speaking, foods and nutrition, safety, or photography. Other competitive activities include judging contests, dress revues, and talent contests. Some examples of noncompetitive educational activities are camps, citizenship short courses, educational tours, and leadership training.

Program Planning Meetings. The term "program planning" usually refers to a list of events necessary for holding a particular event, while "program" may mean the subject-matter emphasis and goals identified for Extension's attention over a given period of time.

Planning meetings may involve a committee of volunteer leaders, the membership of an organized group, or only those in attendance at a public meeting. County planning meetings are likely to have membership selected to represent geographical areas, segments of the population, or subject-matter interests. They may deal with a specific subject or a program area such as youth work. The problems discussed may relate to the entire county, segments of the population, or certain program areas. All planning meetings have the following characteristics in common:

1. The identification of problems needing attention (the situation and the underlying issues, factual information on the prevalence, seriousness, and causes of the problems)
2. The determination of the most practical solution to each problem (the local experience, research findings about the problems)
3. The outlining of a plan of action (the goals, the actions to be taken, by whom, when and where)
4. The evaluation of the program or plan (progress made and accomplishments)

Program planning meetings arouse interest, increase understanding, and contribute to educational growth through the involvement and participation of learners in helping to decide what should be taught and how the teaching should be conducted. Extension uses planning committees extensively. County advisory committees, foundations, and organizational leadership committees give guidance to the Extension program.

Organized Club Programs

Extension has worked through organized club programs with its clientele since the beginning of the organization. While the total family approach was taken, the members of the family soon became segmented in corn clubs for the boys, tomato clubs for the girls, home demonstration clubs for the women, and farm bureaus for the men. (These organizations had various names in different regions of the United States.) Today, segments of these club organizations still exist and represent a major effort of Extension educational efforts in some states. The two most commonly found organized club efforts are the Family Community Education Clubs and the 4-H program.

The 4-H Club. The 4-H program aims to provide educational training for youth through project work, leadership and citizenship programs, and numerous educational activities. The teaching approach is learn-by-doing, sometimes called experiential education, where members develop knowledge, skills, and attitudes needed for their own personal development and leadership.

The 4-H Club is an important vehicle for teaching in Extension work. 4-H members are enrolled in one of several organizational clubs—the community club, school club, project club, project group, special interest group, and teen leader club. The organization of each club differs according to elected officers, adult volunteer leaders, and how individual members are enrolled in selected projects.

The educational base for 4-H club work is the project. A *project* is a given course of study (a curriculum). The member uses project books on the specific subject to guide him or her through self-directed study. These projects offer a wide range of subjects categorized into the major areas of animal science, plant science, science and technology, environmental education and earth science, natural resources, citizenship and civic education, personal development, health and lifestyle education, consumer and family science, and communications and the expressive arts. Project books are available at little or no cost to the 4-H member. One or more projects are required of all 4-H members each year as they are encouraged to "learn-by-doing." Activities, which may be competitive or noncompetitive, are an important part of the 4-H Club member's work. These were explained earlier.

A community-based club has a calendar of activities planned for the year. The club generally meets monthly with business, education, and recreational components to each meeting. The learn-by-doing philosophy extends well beyond specific projects and educational programs. Club officers gain the opportunity to learn parliamentary procedure, teamwork, planning and organizational skills as they conduct club business. Project talks and demonstrations are the part of the educational program where members acquire public speaking confidence. Socialization skills are an important goal of the recreational component. Adult leaders oversee the local club by working with the officer team. Several leadership roles are learned as a result of being a 4-H leader also. These roles include organizational leader, project leader, or activity leader.

The 4-H Club is part of a larger system that includes the 4-H specialists of the state Cooperative Extension Service, county staff, leader and officer associations, USDA, and the National 4-H Council.

Family Community Education. The overall purpose of the home economics program is to help families improve their social and economic well-being. The role of the home economist is to provide a variety of learning experiences in family living education. A wide array of audiences targeted for these programs include young couples, parents, low-income families, senior citizens, and working mothers.

Local clubs are organized at the community level and have elected officers and educational leaders. These leaders have the responsibility to participate in family living training programs and teach this information to local club members. Clubs have an educational lesson at each monthly meeting along with business and recreation.

Leader Training Meetings. The leader training meeting is an educational lesson provided for the 4-H and FCE club leader in a given subject-matter area or on some aspect of leadership. These leaders use this information for their own personal development and for the benefit of the club membership. For example, the FCE leaders in health education may be trained on the benefits of fiber in the diet. The leaders can incorporate these principles into their daily lives but they are also responsible for teaching the same lesson to the local club members during the club meeting. This method of "train the trainer" helps to extend the educational efforts of Extension agents throughout the county.

The organized club programs in 4-H and FCE require continual training of the leaders of the club programs. Leadership training may focus on the duties of officers, program planning, conflict resolution, teamwork, and other relevant topics for groups. Many counties also have 4-H teen leader clubs that require considerable time in training. Most 4-H leaders and FCE club programs have county councils that operate as a separate leadership group within the county. Extension professionals spend a great deal of time recruiting, training, and maintaining the club programs.

Group Meetings by Distance Education

Meetings and demonstrations are a mainstay of Extension's educational program delivery. Educational programs traditionally have been enhanced through videotapes, slide shows, flip charts, and other demonstration materials. This system has served Extension well; however, audio and video teleconferencing can now expand the local meeting to a statewide audience. Teleconferencing is expensive, but when costs are allocated to the number of people reached, it is an efficient, effective means of delivery for the Extension Service. Video teleconferencing has helped specialists deliver the same educational program to hundreds of interested clients on a regional or statewide basis. Some advantages and disadvantages of group teaching methods are listed in figure 7–6.

ADVANTAGES	DISADVANTAGES
Is adaptable to the learning styles of many people	Requires considerable organization and transport of materials and equipment to the meeting locations
Stimulates action as the learner is involved in seeing, hearing, discussing, and participating in the group learning process	Requires a certain amount of showmanship to be successful
Builds confidence in the Extension educator if teaching is performed skillfully	Requires professionals to be effective in public speaking and presentation skills
Lends itself to repeated use or demonstrations by local leaders	Requires a knowledge of a variety of teaching techniques to be effective
Reaches a larger number of people	Limits meetings to certain locations due to the size of the audience
Is adaptable to practically all subject matter	Requires considerable investment in equipment
Recognizes the basic need of individuals to have social contact	Requires flexibility of scheduling to accommodate audience needs and accessibility
Is relatively low cost	Creates difficult teaching situations due to the diversity of audience interests and needs

Figure 7–6 Advantages and disadvantages of group teaching methods

MASS-MEDIA METHODS

Use of the mass media enables Extension educators to greatly increase their teaching efficiency. News stories, circular letters, correspondence courses, radio, television, video, publications, and exhibits provide helpful reinforcement for those contacted personally or through groups. Mass-media methods also make possible the dissemination of information to a much larger and different clientele, especially those with computers. Even though the intensity of the teaching contact through the mass media is less, this is offset by the large number of people reached and the low cost per unit of coverage.

News Stories

The function of the news story in Extension teaching is primarily one of expanding coverage. It is the chief means of getting information about Extension activities and research-proven practices to the clientele who are not contacted individually, do not attend meetings, or do not participate in other Extension activities. A meeting may have only a few in attendance, but the well-written news article giving specific directions for solving a problem will be read by a great number of people.

The Extension news piece may be a feature story advocating the practices of a result demonstration or an article written with information of seasonal interest. Some articles are written by the subject-matter specialist for reuse at the local level while other articles are weekly columns written by the local agent.

The importance of the news story in Extension teaching is emphasized when cost is considered. News stories and radio broadcasts are the least expensive teaching methods when the rate of adoption is considered. Confidence in the reliability of newspapers and illiteracy among certain segments of the population are about the only limitations to the general use of the news story in Extension teaching programs.

Newspapers are adopting electronic technologies to reduce costs, speed the delivery of news, and cut production time. Stories, layout, and composition are received in digital format, edited, and transmitted via satellite to distant printing operations for local delivery. Extension delivery of materials to newspapers via electronic media will continue to grow.

Radio

Radio is the most widely accessible of all the mass media. More than any other medium, radio has the ability to disseminate information to the largest number of people in the shortest time. It is unparalleled as a means of getting emergency or timely information to local people, due to the presence of radios in almost all homes and vehicles. Extension educators can use the radio to disseminate information on market conditions, to inform the public regarding the activities of the Extension Service, to advertise meetings, tours, field demonstrations, and other scheduled activities, and to teach improved production practices. Radio listeners are often encouraged to request Extension literature that provides more detail and may be kept for reference use.

Agents use radio in a variety of ways. They may have a daily program, a weekly talk show, or a call-in program to answer specific questions. Subject-matter specialists may also participate in call-in programs or talk shows. Radio listeners are eager to learn information about situations that are currently affecting their lives. Follow-up information can be obtained from a bulletin, office call, or other contact.

Radio is also being made more efficient by use of electronic technology. Radio stations receive news releases via electronic transmissions over dial-up computer networks and satellite links. Specialized local radio stations are serving selected population groups in rural America, opening new opportunities for Extension to use radio for information delivery and dial-in talk shows with Extension experts. Radio stations are targeting listening populations according to language, age, and income demographics, making it easier for Extension to reach selected audiences.

Television

Television has become a popular teaching tool especially in urban areas. It is more personal than the radio since the audience can both see and hear the professional in a simulated face-to-face situation. With television, the Extension educator can give a "how-to-do-it" method demonstration and reach an audience many times larger than the attendance at a meeting. Close-up pictures may even make it possible for the viewer to see key procedures more clearly than would be possible at a traditional method demonstration. On television, the short talk can be made more effective with visual aids. Videos, slides, and other visuals can be incorporated by television production personnel to better illustrate the before and after, the right or wrong, or the step-by-step process.

Some colleges and state Extension Services operate their own broadcast studios, but most depend upon the educational networks in the state, the local cable channels, or commercial television stations. The availability of television air time is regulated differently from that of the radio stations. Agents in cities with local stations may have a portion of an early morning program, a segment of a noon show, or assist with weekly special-interest segments.

Special USDA efforts to use television as an educational tool include the "Mulligan Stew" television series, which was broadcast nationally with local agents delivering educational instruction about proper nutrition.

Computer-Aided Instruction. The broad category of television also includes many electronic technologies whose common information display mechanism is the television screen. Such technologies include videotape and tape players, videodisc storage and players, computer-generated text and graphic images, and computer-controlled videodisc players. The videotape is a ready-made vehicle for delivery of Extension educational programs. More advanced transmission media for video include satellite transponders, microwave transmission stations, and fiber-optic strands. Many states have been using satellite programming for a few years, either with in-state systems or as a partner of the agricultural satellite consortium. This technology allows for the use of expert speakers in distant locations and in multiple locations. While satellite programming is very expensive in the set-up stages, its impact is far reaching and it has created a new direction for cost-sharing among states and expert specialties.

Computer-aided instruction is used in a limited manner presently in most Extension efforts. A programmed computer-controlled videodisc can deliver educational programs and information with motion, sound, and graphics. Segments of a videodisc can be selected and presented to the student by the computer, making low-cost individualized instruction possible. The computer can be programmed to match the learning pace of the student and reinforce materials if review is needed. Videodisc training materials are now being used to train technicians. This videodisc technology can be adapted to certain educational programs for Extension clientele. This area will see expanded use in the future.

Exhibits

While an exhibit can be a visual aid to strengthen an oral presentation, it can also be a primary means of disseminating information. The most common use of the exhibit is either a window display or an educational exhibit. At the core of many 4-H achievement days are crops, livestock, or other exhibits produced in connection with the 4-H project.

Extension exhibits may be a helpful means of acquainting the general public with Extension work and what it accomplishes. A promotional exhibit may explain the benefits of belonging to a 4-H or FCE club or promote its community activities. The educational exhibit seeks to influence the viewing audience to adopt a better practice or to create an awareness of a community issue or a better standard.

The principles of design are important to creating an effective exhibit. These include the use of color, line, balance, lettering, and movement. Even though they may be much less effective than other, more direct means of influencing practices, exhibits still have a place in creating an awareness of a new technology and fostering good will for Extension on the part of county officials, fair boards, businesses, or other clientele.

Publications

Extension publications are found in many shapes, sizes, and formats. Bulletins, fact sheets, and leaflets continue to play an important role in the present-day system of mass communication available to Extension educators. Publications reinforce other methods of reaching and influencing people. They are distributed in connection with office calls, farm or home visits, and Extension meetings. They are a convenient way to answer requests for information received by letter or phone. Publications amplify and reinforce the news story, radio show, and television broadcast. Guides, pamphlets, brochures, newsletters, and other forms of printed material serve Extension's need for information delivery and support of educational programs. Print materials are better used to supplement other teaching methods than for initiating the teaching process.

The use of digital text (i.e., word processors and computer communications) permits Extension to provide the most current information in its publications. High-quality laser printers and graphic generators are changing the way publications are delivered to county Extension offices. Laser printing and data transmission technology permit quick delivery of a document without the usual cost of duplicating and printing. Electronic technology is changing the need to maintain large publications inventories to satisfy the demand for informational and educational pamphlets and brochures in a timely manner.

Newsletters

Newsletters (circular letters) can be used to stimulate interest in a subject, give timely subject-matter information, announce meetings, obtain information through questionnaires and maintain interest and cooperation of 4-H members, local leaders, and other program cooperators.

Subject-matter specialists often contribute to newsletters. The specialist may prepare a single circular letter or a series of letters relating to a particular subject which can be readily adapted to the local situation. These letters are used to supply subscribers with timely information. An important aspect of the local agent's job is maintaining mailing lists with which specific groups can be targeted for information presented in a particular circular letter.

With the use of electronic publishing software, today's circular letters can be made attractive with a variety of type fonts, layouts, graphic styles, and illustrations that have visual appeal. Electronic technologies and data transmission have automated many publishing functions. Future textual information may be delivered directly to the consumer via computer display, although the cost of high-speed modems, Internet connections, and computers with graphics capabilities currently limit client and agent access to electronically published materials.

Telephone Dial Access

The telephone has become an increasingly important tool in the typical Extension office. In addition to the many uses listed earlier in individual methods, the telephone is used as a dial-in access to recorded messages as an economical way to deliver information to a large number of people. Message systems that convert text to voice and deliver information directly to clients will soon be commonplace. Using telephone push buttons to retrieve problem-solving information, along with newer information display capabilities, will expand the usefulness of the telephone.

Telephone lines can deliver a wide variety of digital data to the client's home. Some data will be processed by the home computer, some printed as news, and some displayed as video programs on television. The telephone instrument will have new roles to play in future Extension delivery systems. The use of the "interactive" classroom that utilizes the telephone line is still in the experimental stages for Extension audiences, but it is sure to be one method of reaching large numbers of people in the future.

Technological advances combining radio, computers, and digital voice transmission are improving the capability of mobile radio telephones in vehicles. Cellular radio technology allows many mobile radio telephones to share the same frequency in a region. Cellular radio will make possible the transmission of digital computerized information to portable computers carried in an Extension educator's car.

Computers

Personal computers are playing an increasingly important role in the information society. The computer can switch, process, and convert digital information streams from one medium to another. It converts text to voice and stores voice messages. The computer can also digitize and store video signals. The computer is at the center of electronic publishing, video text, photo typesetting, electronic mail, and informational database retrieval systems. Computer networks will serve to deliver information, educational programs, and problem-solving services to clients more and more in the future. Many Extension offices have access to the Internet where information on a variety of subjects can be shared among states and countries.

Computer programming research is leading to the development of computers that serve as expert systems. In this role, computers can emulate the logic processes that a subject-matter expert might use to diagnose and solve problems. For example, an expert system can be used to diagnose a plant disease or nutritional deficiencies. While expert systems are not in common use today, experiments are being conducted that may lead to new delivery of information in the future. Some considerations of the use of mass-media teaching methods are given in figure 7–7.

Aids to Teaching

Educational aids are important to the educational process. Used in conjunction with various methods of teaching and learning, educational aids serve to extend the effectiveness of methods and techniques, but they cannot teach by themselves. These aids can include:

1. Illustrative devices such as films, videotapes, audiotapes, slides, overhead transparencies, charts

ADVANTAGES	DISADVANTAGES
Reaches a large number of people at multiple locations simultaneously	Is more expensive than other methods
Reaches those who might not otherwise seek information from Extension	Requires constant revision to stay current
Is a timely source of information because of the frequency and regularity with which information can be delivered	Is limited as a teaching tool for audiences who are nonnative speaking or cannot read
Builds confidence in the local program and university recommendations	Requires training and skills to be effective in written and oral presentations. Some technology requires technical staff assistance
Creates an awareness of problems, issues, or major points	Is ineffective when an editor or producer destroys the intended message or teaching value
Reaches people quickly	Loses effectiveness when educator is not professional in appearance or with presentation techniques
Is adaptable to a wide range of audiences and subject matter information	Is normally broadcast or printed at the convenience of the media
Serves as an effective supplement and reinforcement of other teaching activities	Loses out to entertainment radio and television productions
Lends itself to being read or viewed at learner's convenience	Requires an extensive investment in equipment and network access
Builds an audience of sustained readers, listeners, or viewers	Requires considerable production time for most mass media
Processes or steps that require extended periods of time can be telescoped into a few minutes using video	Time and schedule coordination required for teleconferencing

Figure 7–7 Advantages and disadvantages of mass-media teaching methods

2. Extending devises such as telelecture systems, radio, television, cable television
3. Environmental devices such as seating arrangements and room lighting
4. Manipulative devices such as working models, actual machines in a laboratory setting, simulation trainers

Educational teaching aids can enrich the learning experience greatly, but they must be carefully selected for suitability of the information and to the audience. Aids should be selected after the learner objectives are established and the learning experiences are designed. Learning experiences should never be planned around the media available. It is possible to use a combination of these instructional devices to provide variety and reinforce communication. A good instructor knows when and where to use instructional aids and never tries to substitute them for good teaching (Waldron and Moore 1991).

Electronic Aids

While computers and electronic transmittal systems can greatly change teaching methods in the future, it must be recognized that these methods will only be successful if the information provided is relevant. A videotext system for information on the weather, markets, public policy and technology, and a "stand alone" home computer for the storage and retrieval of on-farm information will be used more and more by clients as an aid to decision making. Commercial vendors as well as the land-grant researchers will establish computerized electronic information retrieval and transmittal systems. Extension personnel will need to monitor the information to ensure that its system provides the best system based upon scientific or economic fact.

The effect of the wide-scale use of videotext for the transmission of information in some branches of Extension work is fairly obvious, especially in the work of the subject-matter specialist. Electronic data relating to livestock and crop production have affected the approach and effectiveness of Extension programs in enterprise management. Some commercial information retrieval networks can be used at very moderate costs to provide assess to many libraries of technical information. The computer can very quickly sort out the specific information and provide a bibliography of sources for those users who desire more comprehensive coverage. The availability of such programs will undoubtedly modify the nature of the service currently rendered by Extension personnel.

There is little doubt that the role of Extension will continue to change in the future. Computers will become more commonplace in local Extension offices where they can be observed by clients and monitored by Extension personnel. Large centralized computer programs will also be accessed frequently from those offices. Extension agents will need to be as proficient in helping producers get the best results from their computers as they are in helping them get the best out of the farm inputs such as fertilizers and chemicals. The initial costs for subscribing to such networks may be high but as more subscribers sign up, the cost will diminish.

SELECTION OF TEACHING METHODS—TECHNOLOGICAL IMPACT

The selection of the teaching method or learning experience is one of the most important roles of the Extension educator. To select learning experiences wisely, it is absolutely necessary to be aware of the knowledge and skills of the clientele. Many factors influence the decision of which method or combination of methods are to be used in an educational program. One of the current considerations is how to use the newer forms of technology wisely. Developments with interactive broadcasts are replacing the concept of a local meeting with the agent in charge. Clientele with access to the Internet are bypassing the specialist and local researcher as the authority in a given subject-matter area. Decision makers must consider how to make investments in these new teaching technologies both effective and efficient. The initial costs of the technological support is high, but the cost per client served is likely to decline as advances continue. These newer technologies embrace the three functional approaches to teaching: problem solving, information delivery, and educational program delivery that have been the mainstay of Extension work for many years. The relationship of these functions to audience size and type is presented in figure 7–8.

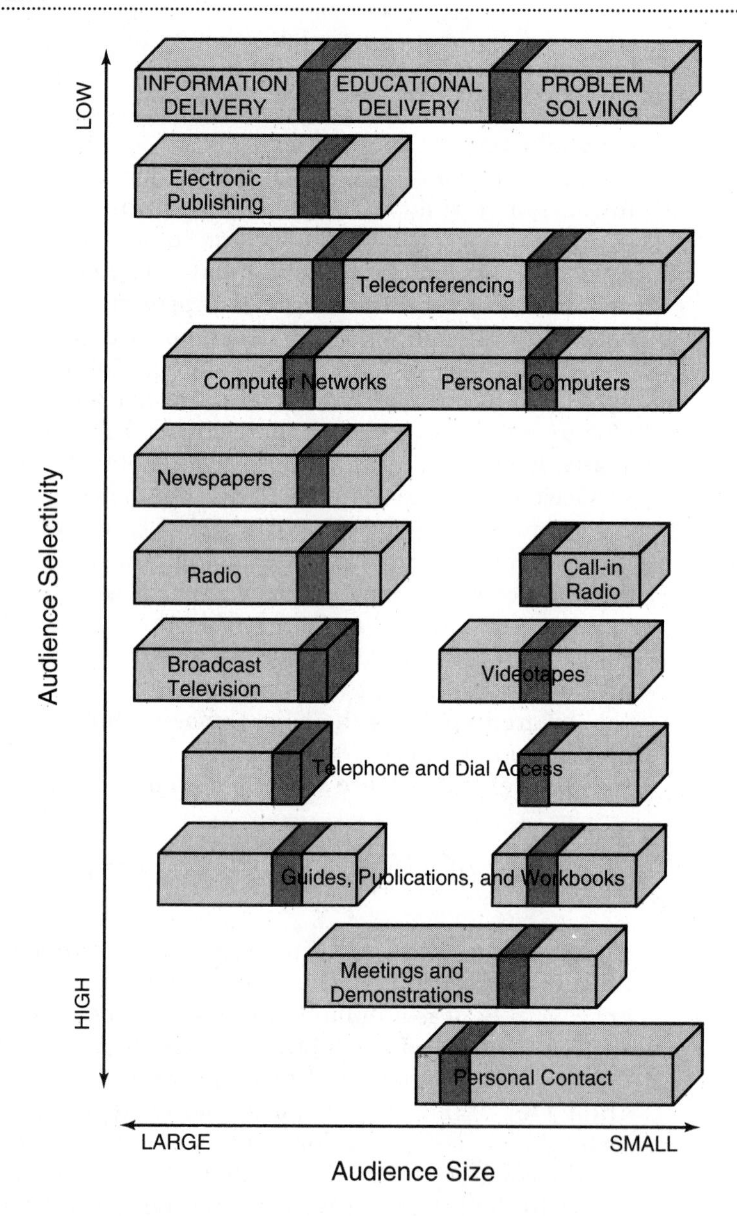

Figure 7-8 ***The relationship of individual, group, and media technologies to the three functions of Extension work, Hussey 1985 p. 163.***

The use of these technologies will impact Extension's educational efforts as they enhance the talents and expertise of the Extension educator, improve services to the traditional clientele, and reach a new clientele who require support and service from an information-rich land-grant university (Hussey 1985). The question is whether Extension will become more an information giver or a problem solver. History will provide the answer.

SUMMARY

Extension educators have used a variety of teaching approaches to help people help themselves. The individual contact has been the foundation of success of the organization. As society changed, Extension added group, community, and mass-media methods to its arsenal of teaching approaches. Most of the success, however, has been the result of the creativity of the Extension educator who applies theory to practice and uses many different approaches to teaching. These educators have continually produced quality programs by understanding the needs and interests of their clientele while reflecting on the past and anticipating the future.

DISCUSSION QUESTIONS

1. At what time are teaching methods planned in the Extension education process? Why not plan programs after the teaching methods are selected?
2. Discuss when it would be more advantageous to use an individual method of instruction, a group method, and a mass-media method.
3. How should each of the following characteristics of learners influence the selection of a specific teaching method: Age? Educational level? Income? Facility location? Rural or urban setting? Time of the year? Subject matter to be taught?
4. Rank each category of teaching method by cost effectiveness and then by teaching-learning effectiveness. When should efficiency be considered over effectiveness?
5. How has electronic technology impacted Extension's teaching methods?
6. How will access to the Internet, on-line data services, commercial marketing services, and other services impact the future of Extension's teaching methods?

REFERENCES

Blackburn, D. J., ed. 1994. *Extension handbook: processes and practices.* Toronto, ONT.: Thomson Educational Publishing.

Hancock, J. 1992. *Agricultural demonstrations.* Lexington, KY: Cooperative Extension Service.

Hussey, A. G. 1985. *Electronic technology: impact on Extension delivery systems.* Extension Committee on Organization and Policy Electronic Technology Task Force Report. University Park, PA: The Pennsylvania State University.

Kelsey, L. D., and Hearne, C. C. 1955. *Cooperative Extension work.* Ithaca, NY: Comstock Publishing.

Leagons, J. P. 1963. *A concept of the Extension education process.* Monograph. Ithaca, NY: Cornell University.

Sanders, H. C. 1972. *Instruction in the Cooperative Extension Service.* Baton Rouge, LA: Louisiana State University and Agricultural and Mechanical College. .

Sanders, H. C., et al. 1966. *The Cooperative Extension Service.* Englewood Cliffs, NJ: Prentice-Hall.

Waldron, M. W., and Moore, G. A. B. 1991. *Helping adults learn.* Toronto, ONT.: Thomson Educational Publishing.

Wilson, M. C., and Gallup, G. 1955. *Extension teaching methods.* Extension Service Circular 495. Washington, D.C.: Federal Extension Service, United States Department of Agriculture.

C H A P T E R

"Every day is a fresh beginning . . . take heart with the day, and begin again."

—SUSAN COOLIDGE

Evaluating Extension Programs

KEY CONCEPTS

- Elements of evaluation
- Reasons to evaluate
- How to evaluate

INTRODUCTION

Informal evaluations, such as deciding which movie to see and which shirt to wear, are a part of daily living. In much the same way, Extension educators evaluate programs informally by listening to casual comments and asking councils for advice. But Extension legislation requires formal program evaluations, following systematic procedures that can be verified and repeated. Formal evaluations are also important as Extension faces public scrutiny of its programs. An evaluation report is visible evidence to others of the worth of a program.

Definition of Evaluation

A simple definition of evaluation is the systematic process of determining the worth of a person, product, or program. Determining the worth of Extension programs is a continuous process that is essential at all three stages of Extension's educational program efforts: planning, design and implementation, and evaluation. An evaluation at the design stage of a program might answer a question such as; "What might solve this community's problem?" An evaluation during the delivery of a program would ask "How are things going?" And an evaluation at the documentation stage of a program searches to answer the question "Should we try this program again?"

Preliminary Questions

In formal program evaluations, there are a number of decisions to make. First of all, what should be evaluated? This question is not as simple as it first appears. Describing specifically what needs evaluating is an important first step. Once that has been determined, the following questions need to be answered:

- Why should it be evaluated?
- Who wants to know?
- What resources of time, people, and money are available to do the evaluation?
- What is the best method to use?
- How should the results be reported?

Evaluation Components

There are many different ways to evaluate, but they all include two elements: (1) gathering the evidence and (2) comparing the evidence with criteria, see figure 8–1.

Gathering the evidence consists of collecting information about the program, product, or person. The evidence may be in numbers or in words and is called the data. Comparison consists of making decisions about the meaning of the evidence. It is the process of judging the evidence against a set of criteria. A good set of objectives can serve as criteria, in which case the evaluation consists of determining whether or not the program has met the objectives. Writing program objectives was discussed in Chapter 5.

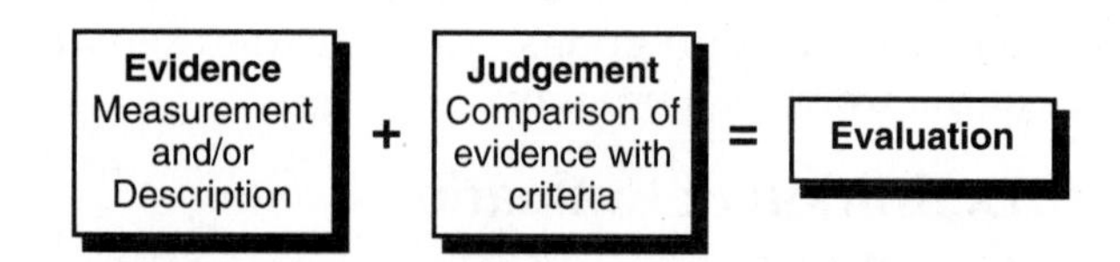

Figure 8–1 ***The Elements of Evaluation. Julia Gamon, Department of Agricultural Education & Studies, Iowa State University, Ames.***

REASONS TO EVALUATE

Reasons to evaluate fit into two main categories: either (1) to prove something (accountability) or (2) to improve something. Evaluations whose purpose is accountability are called **summative evaluations**, while evaluations for improvement reasons are **formative evaluations**. Formative evaluations yield results useful to improve programs, whereas summative evaluations yield results useful for making decisions on the continuation of programs. Seldom are evaluations strictly formative or strictly summative; their purposes usually overlap. Even when required to evaluate a program to complete a mandated accountability report, the findings should be useful for program improvement.

Reasons To Use Summative Evaluation

Collect evidence for reports

Document achievement of objectives

Justify expenditures

Assess program impact and outcomes

Document that new practices were adopted

Increase visibility of programs

Promote public relations

Provide evidence of program accomplishments to voters, private donors, youth, families, volunteers, and Extension personnel

Ensure that client expectations are met

Measure economic benefit to participants, communities, and society at large

Reasons To Use Formative Evaluation

Find out what works or doesn't work with a particular audience and situation

Help educators grow professionally

Increase efficiency

Provide personal satisfaction

Reasons To Use Both Formative and Summative Evaluation

Provide a basis for decisions on beginning, changing, and stopping programs

Give a rationale for personnel changes

Set priorities

Give direction to programs

Establish vision

Allocate and reallocate resources

Formative evaluations often occur while programs are in progress and are more likely to be something educators do on an informal basis. Summative evaluations are likely to occur after programs are over and to use outside evaluators. Formative evaluation is really an "early-warning summative" (Scriven 1993), while there is still time to make improvements and to identify what needs changing.

Reasons Not To Evaluate

Some programs should not be evaluated. Smith (1989) suggests doing an evaluability assessment of programs or potential programs to determine if an evaluation is an appropriate investment of time and resources. The following are reasons not to evaluate:

When no one is going to use the information

When the program is intended to be used one time only

When the program is just getting started or is going through a period of change (a formative evaluation could be done, but not a summative one)

When the purpose of the program is vague

When stakeholders are unclear about what should be evaluated

Accountability Mandates

There have been three accountability mandates for Extension issued by the federal government. Extension's first accountability mandate was a part of the 1914 Smith-Lever Act, which called for "a full and detailed report of its operations." In 1977, Section 1459 of the 1977 Food and Agriculture Act ushered in a new era of evaluation activity when it directed the Secretary of Agriculture to transmit to Congress "an evaluation of the economic and social consequences of . . . programs" (Congressional Record 1977). In 1993, the Government Performance and Results Act required strategic plans and numerical assessment of outcomes as a part of its measurement of performance of governmental organizations. Chapter 11 discusses Extension's response to this legislation in greater detail. Figure 8–2 displays a model developed by the U.S. Treasury Department for performance measurement. The phases of the model—strategic planning, creating indicators, developing a data measurement system, refining performance measures, and integrating with management processes—match basic first steps in program evaluation.

MANAGING AN EVALUATION

Managing an evaluation is similar to managing a program in that it requires planning the use of time, people, and resources. The old saying,

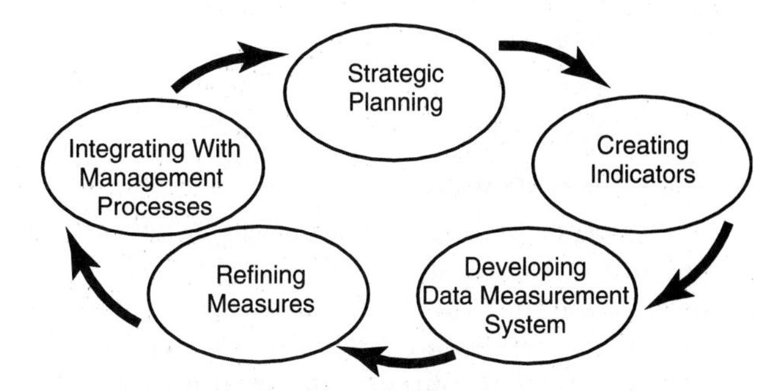

Figure 8–2 ***A Model Performance Measurement Cycle for Governmental Agencies. 1993. U.S. Treasury Department.***

"Six serving men have I: Who, What, Where, When, How, and Why" applies to managing evaluations. The who, what, and why help to identify the stakeholders, to find out what they need from the evaluation, and why they need it. Where, when, and how are best defined by a series of steps and a time line that identifies who will do what, when. An efficient manager will delegate some of the responsibility and authority for the evaluation to others, such as support staff, volunteers, or other agencies.

More and more often, Extension staff find themselves involved in writing proposals to secure grants for funds to underwrite programs. A guide for estimating the cost of the evaluation component is 10 percent of the cost of the program. About 70 percent of evaluation costs typically go toward personnel expenses. The grant will usually request a time line as well as a budget.

The Role of Stakeholders in Evaluation

Strategic planning includes taking into consideration external factors that impact a program's future. Included in the consideration should be the stakeholders—people with an interest in a program. Taxpayers, advisory council members, supervisors, and upper-level administrators are all stakeholders who are interested in proof of a program's impact. Other stakeholders are the participants in the program, those receiving the benefits of the program, and those helping deliver the program.

Participatory evaluation theory suggests that when stakeholders take a more active part in planning evaluations, collecting information, and reporting results, the results of an evaluation are more likely to be used. An often-overlooked resource in evaluation is volunteers. Some studies have used stakeholders as volunteers to do telephone surveys, analyze the findings, and report the results to councils and committees.

Evaluation Standards

Managing an evaluation implies concern for its quality. Four criteria for measuring the worth of an evaluation are (Sanders 1994):

- Is it useful?
- Is it feasible?
- Is it proper?
- Is it accurate?

Whether or not an evaluation is **useful** may hinge on the timing and the appropriateness of the questions asked. **Feasibility** considers such issues as whether the evaluation can be done in time to use the results, its cost effectiveness, and whether it is manageable in terms of personnel and technical resources. How **proper** an evaluation is relates to ethics—whether the questions asked are appropriate to the topic or are seeking private information unnecessarily. Questions concerning an individual's age or income are often sensitive issues. A proper evaluation asks only for information that will be used. Generally, preserving confidentiality results in more trustworthy data. If an evaluation summary reveals an individual's response in an identifiable way or the evaluator provides individual responses to a supervisor without prior permission, confidentiality has been violated. Evaluation data should be summarized so that only group responses are reported to ensure confidentiality.

The **accuracy** of an evaluation rests on choosing samples randomly, making follow-up requests to obtain representative returns, not trying to apply the results to groups other than the original, and carefully measuring, analyzing, and reporting. Even if the sample is randomly chosen, if fewer than 50 percent of the sample group return a questionnaire, their views cannot legitimately represent all those who did not respond.

STEPS IN EVALUATION

There are five major steps in the evaluation process. The first two are planning steps and the last are action ones. The first step in evaluation is selecting what to evaluate. This entails describing specifically the object of the evaluation and identifying the purpose; is it "to prove" or "to improve?" The next step is to plan the strategy for how the evaluation will be conducted. Often an evaluation consists of determining to what extent a program's objectives have been met. After the evaluation is planned, the final three steps are collecting the evidence, analyzing it, and reporting the results.

A word of caution: Sometimes the objectives are not appropriate, or unanticipated effects become important, or the situation changes. Sometimes it is better to first observe the program and then develop criteria to measure its worth.

In all cases, it is important to involve the stakeholders in as many of the steps as possible. The chart in figure 8–3 can help in planning the evaluation.

EVALUATION STEPS	STAKEHOLDERS							
	Self	Part	CoWrk	AdvG	Spec	Adm	Leg	Fund	Pub/M
1. Select what to evaluate									
2. Identify purpose									
3. Study objectives									
4. Choose level of impact									
5. Decide when, how, who									
6. Find forms to use									
7. Test or pilot									
8. Analyze data									
9. Summarize findings									
10. Write reports									
11. Distribute reports									
12. Evaluate the evaluation									

Self = myself
Part = participants in the program
CoWrk = coworkers and colleagues
AdvG = advisory groups
Spec = district and state Extension specialists
Adm = Extension administrators
Leg = legislators and commissioners
Fund = funding agencies and sponsors
Pub/M = public and media

Figure 8–3 ***Involvement of Stakeholders in Evaluation Steps. Julia Gamon, Department of Agricultural Education & Studies, Iowa State University, Ames.***

OBJECTIVES-BASED EVALUATIONS

Extension education uses the objectives-based approach in planning and evaluation, an approach similar to that used by formal education in school systems. Ralph Tyler, an eminent educator, whose work came into prominence about the same time as the beginning of Extension education, was a strong proponent of the use of objectives.

Role of Objectives

Although the most important role of objectives is to provide direction for instruction, of next importance is serving as criteria for evaluation. Objectives also help communicate expectations, and they provide a framework for reports.

Hierarchy of Evidence

Extension typically uses a specific form of the objectives-based approach, Bennett's Hierarchy (Bennett 1975) for evaluation of programs. When using the Hierarchy, specific objectives should be written according to the level of impact desired.

Questions to Ask. The following questions are addressed at each level of the Hierarchy.

Level 1: How many staff, how much time, how much money?

Level 2: What kinds of meetings and other delivery methods?

Level 3: Who and how many participants? How often and how long?

Level 4: Satisfaction of participants?

Level 5: Changes in participants' knowledge, attitudes, skills, and aspirations?

Level 6: Changes in practices used by participants?

Level 7: Changes in the lives of participants and the society in which they live?

Measuring the Level of Impact. Impacts at higher levels of the Hierarchy demonstrate greater accountability than impacts at lower levels. However, the cost of an evaluation in time and money also increases. The lower levels, Levels 1 and 2, measure inputs (through attendance figures, for example), what people deem important by their time and support. The higher levels, Levels 3 through 7, measure effectiveness of programs in terms of impact, such as use of new tillage methods. An example of economic evidence at the top level is how much profit swine farmers have realized from computer enterprise analyses.

Evaluations targeted at the fifth level measure changes in knowledge, attitudes, skills, and aspirations. These changes are often measured using clients' perceptions. Spiegel and Leeds (1992) give some tips on measuring knowledge:

> *First, note whether in your objective you want to measure an* actual increase *in knowledge or a* perceived increase *in knowledge. [For actual knowledge] you must have done a premeasure . . . and a post program measure. If you ask them at the end of the program whether or not their knowledge has increased, [this question] will tell you what participants think they know [but not] what they actually know*
>
> (SPIEGEL & LEEDS 1992, P. 16).

The type of instrument used for the pretest and posttest is important. The same questions should be used for both tests, and they need to be questions that have been proven to provide a valid measure of knowledge.

Creating Indicators

Possible indicators to use in evaluating programs are efficiency, effectiveness, and expectations of clientele. The following are examples of questions for each kind of criteria:

- **Efficiency:** Did we do it right?
  - Were the arrangements satisfactory?
  - Was the time and place convenient?
  - Was it worth the time and effort?
  - How much in time and money of presenters and participants did it cost?
  - Did the benefits outweigh the costs?
  - What would people be willing to pay to do it again?
- **Effectiveness:** Did we do the right thing?
  - Was it what was needed?
  - What has changed as a result of the program?
  - To what extent has it changed?
- **Expectations of clientele:** Did it meet expectations?
  - Were people satisfied?
  - Was it of high quality?
    - -Interesting?
    - -Current information?
  - Was it relevant?
    - -Important topic?
    - -Important information?
    - -At the right level of difficulty?
  - Did it deliver what was promised?
    - -Was it usable?
    - -Answered questions?
    - -Solved problems?
    - -Did participants learn something new?
    - -Did participants learn a better way?

Complex educational programs involving multiple meetings or delivery methods will require multiple measures. There may be different stakeholders and different criteria for different aspects of the program. For example, a pretest and posttest could be used to determine if participants gained knowledge in a nutrition program. A six-month follow-up interview could be used to determine if the participants actually changed food preparation and eating habits based upon the knowledge learned.

Collecting Evidence

If at all possible, use an existing questionnaire that has been tested and has been proven to be valid and reliable. Some states provide end-of-meeting forms, such as the example in figure 8–4.

Evaluations often gather evidence in both numbers and words, just as a test may have both essay and multiple choice questions. Some of us are

Program Topic_____________

Please rate this program on its relevance, its quality, and its usefulness, by circling a number. Your responses will help us meet your expectations. (circle number)

RELEVANCE

	Yes, very much so				No, not at all
1. Did this program deal with important concerns?	5	4	3	2	1
2. Was the content what the program announcements said it would be?	5	4	3	2	1

Comments about the relevance of the program:

QUALITY

	Yes, very much so				No, not at all
3. Was this program					
a. current, up-to-date?	5	4	3	2	1
b. understandable?	5	4	3	2	1
c. presented in an interesting way	5	4	3	2	1

Comments about the relevance of the program:

USEFULNESS

	Yes, very much so				No, not at all
4. Did this program answer your questions?	5	4	3	2	1
5. Did you learn?	5	4	3	2	1
6. Do you intend to use any of the information from this program	5	4	3	2	1

Comments about the relevance of the program:

	Yes, very much so				No, not at all
7. **Overall**, was this program worth your time to attend?	5	4	3	2	1

Comments about the relevance of the program:

Thank you!

Figure 8–4 ***Generic Evaluation Format. Adapted from Smith (1992)*** **Criteria for Excellence** ***College Park, MD. University of Maryland Cooperative Extension Service.***

"numbers people," while others prefer to read or listen to words. **Quantitative data** refers to evaluation evidence that can be expressed in numbers or statistics. **Qualitative data** refers to information that describes program inputs or impacts in words, phrases, or quotes. Together, quantitative and qualitative data support one another to create a more complete picture concerning program inputs and impacts.

Quantitative Data	*Qualitative Data*
Numbers	Words
Measurement	Descriptions
How many	Comment section
Objective	Subjective
Statistics	Case studies

When using a questionnaire that asks people to circle numbers, it's wise to include open-ended questions and spaces for people to write comments as well. This adds qualitative evidence and quotations that can be used to explain and support the numbers.

DATA COLLECTION METHODS

Several possible ways to collect evidence are listed below. They are discussed in more depth in the paragraphs following.

Existing records (Census figures, production records)

Telephone surveys

Mailed questionnaires

End-of-meeting questionnaires

Nominal group process

Focus groups

Concept mapping

Card sorts

Observations

Public hearings

Face-to-face (personnel evaluations, leader interviews)

Most data-gathering techniques can be adapted to collect evidence at any stage of programming: planning, design and implementation, and evaluation. The choice of the technique depends upon the nature of the program and the purpose of the evaluation. However, certain techniques are typically used at particular programming stages. Needs assessments occur at the time when programs are designed. Appropriate methods include the nominal group process, focus groups, card sorts, and concept mapping. Site visits and interviews are often the method of choice at the delivery stage of programming and are used to monitor programs and personnel. After programs are completed, common methods are telephone surveys, end-of-meeting forms, and mailed questionnaires.

Construction of Questionnaires

Questions are a part of all evaluations, at all stages, whether assessing needs, monitoring delivery, or measuring the impact of the program. In some approaches the questions are general and evolve as the evaluation proceeds. However, many approaches use structured questionnaires that contain a series of questions. Sometimes Extension educators need to modify available questionnaires or construct new ones. Useful references on constructing questionnaires are those by Dillman (1978) and Salant and Dillman (1994). All questionnaires should be tested with a sample of the intended audience to find out if the questions actually measure what needs to be measured in a particular situation.

Tips for Constructing Questionnaires. Here are some pointers for choosing or constructing questionnaires:

- Use the same scale for all questions; people find it easier to respond.
- Scales with five or more points allow more insight into participants' responses than yes or no answers.
- When rating people, especially if the purpose is to help people improve, words are less threatening and more encouraging than numbers.
- Put questions about age, gender, and income at the end. A first question needs to be more interesting than "How old are you?"
- Take out any unneeded questions, and shorten, shorten, shorten.
- If a question is asking about two things, divide it into two questions.
- Choose words or phrases with very clear meanings—for example, "Did you increase your net income?" instead of "Did you make more money?"

GROUP TECHNIQUES FOR NEEDS ASSESSMENT AND EVALUATION

Group techniques such as the nominal group process, focus groups, and concept mapping are commonly used to gather evidence for the planning and designing stages of program development, but they may be adapted for evaluation at any stage. All gather evidence in group settings, and all should begin with a clear statement of the task, for example, "What are the most pressing needs of the elderly in our community?"

Nominal Group Process

The nominal group process (Delbecq, Van de Ven, and Gustafson 1975) is so called because it is a group in name only; several of its steps require individuals to work silently. The process moves from individual priorities to a group consensus on the most important items and thus combines individual creativity with group critiques. The nominal group process is particularly useful in formative evaluation to identify needs and to determine group consensus on content or methods to use for a program. Its advantages are that it is suitable for large groups and provides every participant an equal chance to be heard. Its disadvantage is that ownership of ideas gets lost in the process. Even though everyone may agree on the needs, that may not translate into individuals who are willing to provide leadership for action.

Time required:	One to two hours
Room set-up:	Space for 5 to 8 persons in each small group
Materials:	Newsprint, marking pens, paper and pencils for each person.

Procedure:

1. State the task in brief, precise words, and write it out large enough for everyone to see. Divide the participants into small groups and ask for a recorder in each group.
2. Give newsprint and a marking pen to each recorder
3. Each person, working silently and independently, should jot down three to six responses to the task.
4. Within each group, one by one, each person gives one idea (in a few words), and the recorder writes these ideas (without comment or changes) on the newsprint for all to see. Continue around, one idea per person, until all ideas have been listed and numbered.
5. Each small group discusses, clarifies, and ranks its ideas.
6. Bring the small groups together into one large group and ask each group in turn to suggest one idea, starting with their highest ranked one. List these on newsprint by number and proceed until all ideas have been gathered. Ask for explanations and suggestions for combinations of ideas.
7. Individuals rank order their top three preferences, and come up with a small group ranking.
8. Results from each group are orally announced and scores written on the master list.

Focus Groups

Focus groups are small groups of strangers with a common interest who discuss a topic in a systematic manner. Krueger (1994) offers detailed guidelines for conducting focus groups. It is important to follow suggested procedures carefully. For example, the discussion leader needs to be someone who is not connected with the program. Focus groups may be used for formative or summative evaluation. They result in rich, descriptive information.

Features of focus groups are the following:

- Trained discussion leaders
- Well-thought-out questioning route
- Seven to ten participants in each group
- Two to three groups for a topic

- Each group's participants are selected on the basis of a common interest or common characteristic
- Reimbursement for participants
- Discussions lasting one and a-half to two hours
- Discussions tape-recorded and carefully analyzed later
- Enjoyable for all

Concept Mapping

Concept mapping is a technique that employs graphic representations instead of words and numbers. Sometimes called mind mapping, it is useful for drawing out creative responses. Concept mapping works with planning as well as needs assessment and can be done by both individuals and small groups. Researchers have used computer programs to analyze concept mapping results and document the strength of the needs.

The **futures wheel** is a type of concept mapping that has been used in needs assessment in the 4-H program area (Etling and Maloney 1994). In figure 8–5, 4-H project leaders who drew the futures wheel were able to

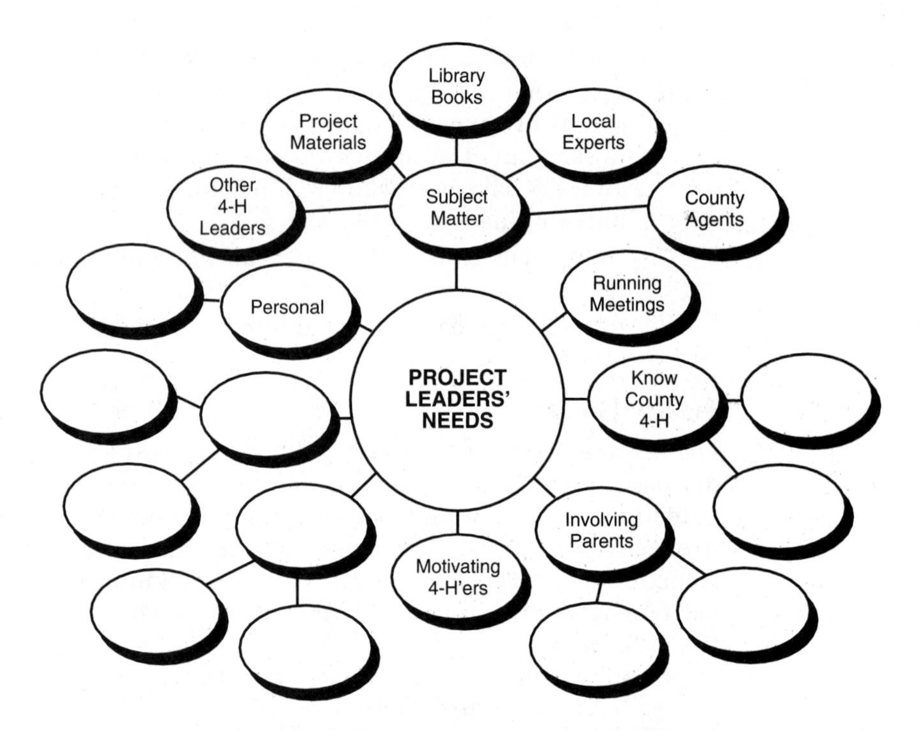

Figure 8–5 *Futures Wheel. Arlen Etling & Thomas Maloney. 1994.* **Needs Assessment for Extension agents & other nonformal educators** *(p.19). Pennsylvania State University, College of Agricultural Sciences, Cooperative Extension.*

generate a number of ideas for places to find subject-matter information. An evaluation of how leaders teach subject matter to 4-H'ers might focus on access to information sources.

Step 1. Draw the wheel on large pieces of newsprint and fill in the center circles and a few of the other circles.

Step 2. Give one of the pieces of newsprint and a marking pen to each group of three to five people. Encourage them to doodle and draw in or around the circles. Colored markers or pencils and people with skills in drawing enhance the process.

Step 3. Hang the maps on the wall.

Step 4. Discuss the maps.

Card Sort

The card sort technique is a prioritizing process that uses duplicate sets of small cards (calling card size). Each card contains a single problem, need, or program effort. Advantages of the card sort technique are that a large number of items (up to fifty) can be prioritized in a short amount of time. Input related to the items on the cards can be generated from a number of sources. Groups of people can do the card sort at meetings or cards can be mailed out to clients.

Step 1. Generate items for the cards. (The items on the cards are generated beforehand by Extension educators, advisory groups, and other stakeholders, although blank cards are often provided for individual use during the card sort.)

Step 2. Give each person a set of cards to separate into piles (three to five piles) according to their priority for programming efforts.

Step 3. Prioritize the cards in the top pile.

Step 4. If the group is small, each person may call out the top entry on the top pile, and these may be written on a chalk board or easel. It may be possible to go around the group again and mark some items as second choices.

Step 5. Or, piles may be marked with Post-it Notes, fastened with paper clips or rubber bands, and collected for later analysis.

Brainstorming

Brainstorming is a group technique often used in the middle of a discussion to assist with problem solving. Brainstorming is the phase that generates ideas; it needs to be followed by analysis and a plan for action. Strengths of brainstorming include the following:

- Generates many ideas in a short length of time
- Encourages creative thinking, such as analogies to animals, different situations

- Ideas build upon each other
- It's fun!

Rule 1. Call out ideas; don't wait to be recognized

Rule 2. Anything goes (the wilder, the better)

Rule 3. No criticizing or judging of ideas

Rule 4. Use others' ideas as springboards for your own: expand, reverse, consolidate.

Brainstorming is useful in formative evaluation to solicit group input for the development of a program or to improve a program while it is in progress. It could also be used in summative evaluation to determine future directions for a program.

Program Reviews

Program reviews typically last several days and include a team of three or more persons external to the program. Team members often include an administrator, a person who works in a program similar to the one under review, and a content specialist. Because program reviews are expensive in terms of the amount of time required by all, they are reserved for major program efforts.

The following steps have been found to be successful:

1. Identification of the purpose of the program review
2. Identification of the stakeholders in the program and arrangements to include them in the process
3. Development of an in-depth description of the program. This part of the process is a valuable one, because it requires those who are working within the program to gather evidence before the visit by the team.
4. Visit site. Prior arrangements need to be made so that the team can see enough people and parts of the program to gain an accurate view.
5. Develop and write reports. Typically the visiting team gives the oral report at the close of the site visit with the written report arriving later.

The Delphi Technique

The Delphi technique (Moore 1987) is a useful tool when experts on a topic are geographically separated. Because it uses a series of mailed questionnaires, starting with a few open-ended questions, it usually takes several months to complete. The results from each questionnaire are used to construct the next questionnaire. It has the advantage of moving a group of experts to consensus without undue pressure from dominant individuals. First developed in the 1950s to make military projections, it is a good forecasting tool. This process is especially useful to design an evaluation instrument. It could also be used as a way to determine programming needs or to determine group consensus about the impact of a program upon its completion.

Program Monitoring

Monitoring during program implementation is important because many times learning activities are not implemented according to the original design. Needed resources such as time, personnel, and materials may be limited or missing (Boone, Pettit, and Safrit 1994). Monitoring serves the following purposes:

- Provides evidence that the program did take place
- Documents whether or not the program served the target audience as intended
- Identifies constraints and variations on the program design

Monitoring involves more than counting the activities of the educator or measuring yields. It is important to consider the motivations that are behind actions of personnel and clients (Murphy and Marchant 1988). Site visits for monitoring purposes may be made by an individual or by a team; in the latter case the visit is called a program review.

Self-monitoring

Although monitoring is usually done by external evaluators, self-monitoring has proven to be an effective technique for evaluating during the course of a program. Self-monitoring empowers individuals and encourages them to continually improve their own performance.

Dube and Martin (1992) suggest that self-monitoring would "improve extension at the delivery point." They made the following observations:

- The autonomy that comes from self-evaluation is energizing and leads to improved performance.
- The individual evaluating his or her own programs and performance can take immediate corrective action.
- There is less need for supervision, thus freeing resources for other activities.
- Individuals can look critically at their programs without outside pressure to inflate results.

Some of the rationale for self-monitoring comes from the Total Quality Management (TQM) movement. According to Deming (1986), only 15 percent of what a person is able to accomplish is under his or her own control; 85 percent of the results of a program are influenced by the organization or outside effects. The conclusion is that monitoring of programs should focus more on the organization rather than passing judgement on the worth of individual staff members.

ANALYZING THE DATA

Once all of the evidence is in, the next step is to organize and analyze it. This is easier if there is a plan for how to do this from the beginning. For example, it is not necessary to ask everyone every question. Random sampling of an appropriate number of people will give accurate results. For some evaluations, purposive sampling is appropriate. Guidelines and formulas for sampling are included in Salant and Dillman (1994).

Some Extension personnel have access to support staff who do all the analyses for them. Others have their own microcomputer programs for statistical and qualitative data analysis. Some ideas for handling survey data include (Norland 1991):

1. If many people (every fifth person, e.g.) are not answering an item, take it out of the analysis.
2. If a survey has only a few responses marked, don't use any of it.
3. Don't guess at what is meant. Don't try to interpret marks between choices.
4. If more than one out of five are circling "NA," not applicable, for an item, don't use that item in the analysis.

A set of frequencies and means may give all that is needed. This will indicate the number for each response level and the average response. But we may want to see if there are relationships between responses and ages of the clientele. Or we may want to find out if there are differences between those who live in town and those on the farm or those who participated and those who did not. These questions require more sophisticated analyses and perhaps the assistance of a statistician.

Averaging Ratings

If the questionnaire is simple with a small number of respondents, the quickest, easiest, way to generate average responses may be to do it by hand. Figure 8–6 contains an example of a cross-hatch method that an office assistant or a volunteer might use.

	SD	D	U	A	SA
The content was relevant to my situation.			~~IIII~~ II	~~IIII~~ IIII	~~IIII~~
(21) (36) (25)	SD = 1	(Strongly Disagree)			
7x3 + 9x4 +5x5 = 82	D = 2	(Disagree)			
7 + 9 + 5 = 21	U = 3	(Undecided)			
82 ÷ 21 = 3.9	A = 4	(Agree)			
	SA = 5	(Strongly Agree)			

Figure 8–6 ***Cross-hatch Analysis. Julia Gamon, Department of Agricultural Education & Studies, Iowa State Univesity, Ames.***

The person who is compiling the data uses a blank questionnaire for a master sheet and goes through each questionnaire, putting marks in the appropriate columns. Then each set of marks is multiplied by the value of that column and the total is divided by the total number of responses. The average response to the example above was 3.9, interpreted as "people agreed that the content was relevant."

Multi-attribute Analysis

Sometimes there are several criteria that need to be considered in an evaluation, and some criteria are much more important than others. Multi-attribute analysis takes into account how people rank the criteria as well as how they rate each item. It is a useful tool for decisions such as hiring personnel, selecting equipment, or choosing a site for an office, see figure 8–7. Steps in the process are:

1. Decide on attributes (criteria) and points. It is a good idea to decide on the total number of points first (e.g., 10). Then reach consensus among stakeholders on how to divide up the total among the various attributes.
2. For each attribute, place item in rank order.
3. For each item, multiply the attribute's points by the rank order number.

1. Decide on points for attributes

Attribute	Points
Client accessibility	3
Cost	3
Space appropriate	3
Staff convenience	1
Total points	10

2. Rank potential office sites. Best choice = 3; Low choice = 1

Site	Client Access $(3)^a$	Cost $(3)^a$	Space $(3)^a$	Staff Conv. $(1)^a$
Fairgrounds	3^r	2^r	3^r	1^r
Inner City	1^r	1^r	1^r	3^r
Outskirts	2^r	3^r	2^r	2^r

3. Multiply attribute points[a] by ranking[r]

Fairgrounds	Inner City	Outskirts
$3^a \times 3^r = 9$	$3^a \times 1^r = 3$	$3^a \times 2^r = 6$
3 x 2 = 6	3 x 1 = 3	3 x 3 = 9
3 x 3 = 9	3 x 1 = 3	3 x 2 = 6
1 x 1 = 1	1 x 3 = 3	1 x 2 = 2
25	12	23

Figure 8–7 ***Multi-attribute Analysis: Selecting an Office Site. Julia Gamon, Department of Agricultural Education & Studies, Iowa State University, Ames.***

4. Sum each item's total score. The highest score corresponds to the best choice. Once stakeholders have agreed on the relative points for each attribute and then ranked each possible choice, they are usually willing to accept the choice with the highest score. In figure 8–7, the fairgrounds turned out to be the best site for a new Extension office.

REPORTING THE RESULTS

An evaluation is not complete until the results are reported. The most common mistake in reporting is "one size fits all," assuming that one report is appropriate for all stakeholders. The next most common mistake is waiting until the end of the evaluation to report. Particularly for large evaluations, progress reports, which may be in the form of telephone or computer messages or brief notes, keep the final results from being a surprise, either pleasant or unpleasant.

Format

The format of the report needs to fit the audience. Possible formats are newspaper, radio, and TV reports for the general public. Other stakeholders may require videotapes, booklets, or long or short written reports. Oral reports by volunteers or youth to committees and councils should also be considered. Presentations may be formal or informal, with or without graphs and charts.

Contents of Reports

Contents of evaluation reports usually include a description of what was evaluated, methods used, summary of what was found, what might be concluded from the findings, and recommendations for action.

An evaluation may generate controversy, anxiety, or just plan apathy. Patton (1987) said that evaluation findings are usually political in nature. There may be rivalries, jealousies, budget battles, power struggles, and external pressures that affect how others react to evaluation reports. Even if the results are what everyone expected, a well-planned and implemented evaluation reduces some of the uncertainty decision makers inevitably face (Sawer 1992).

SUMMARY

Evaluation assesses the value or worth of something or someone. It may be used either to prove (summative) or to improve (formative) the value of a program. It is a continual process taking place throughout the life of a program, and its forms include assessing needs, monitoring activities, and measuring changes occurring as a result of program impact.

The formula for evaluation is: Evidence + Judgement = Evaluation. Criteria useful for judging the worth of a program are efficiency, effectiveness, and whether the program met clients' expectations for relevance, quality, and usefulness.

Extension uses Bennett's Hierarchy of Evidence and objectives to measure each level of desired impact. The evidence may consist of qualitative data (words) or quantitative data (numbers) or both.

Evaluation steps include (1) focusing on the object to be evaluated, (2) planning the strategy, (3) collecting the evidence, (4) analyzing and interpreting the evidence, and (5) reporting the results. An evaluator needs to manage all of these steps, making sure to involve the stakeholders during the process.

A number of strategies may be used to gather evidence for making judgements, including the nominal group process, focus groups, and questionnaires. The choice of strategy depends upon the purpose of the evaluation and the nature of the program.

DISCUSSION QUESTIONS

1. Choose a current Extension program and decide whether formative or summative evaluation would be more appropriate and explain why.
2. How do the accountability mandates from the federal government affect the designs of Extension evaluations?
3. Explain the connection between program planning and evaluation using Bennett's Hierarchy.
4. Assume that an Extension specialist is conducting an evaluation workshop to teach county or parish Extension staff and interested volunteers about how to evaluate a program or event. What points should be emphasized?

REFERENCES

Bennett, C. F. March/April 1975. Up the hierarchy. *Journal of Extension* 13(1):7–12.

Boone, E. J.; Pettit, J., and Safrit, R.D. 1994. Program evaluation in Extension. *Journal of Extension Systems* 10(1):87–121.

Congressional Record. 1977. *Food and agricultural act, Section 1459.* Washington, D.C.: U.S. Congress.

Delbecq, A.; Van de Ven, A.H.; and Gustafson, D. H. 1975. *Group techniques for program planning.* Glenview, IL: Scott, Foresman & Co.

Deming, W. E. 1986. *Out of the crisis.* Cambridge, MA: Massachusetts Institute of Technology.

Dillman, D. A. 1978. *Mail and telephone surveys: The total design method.* New York, NY: John Wiley & Sons.

Dube, M. A., and Martin, R. A. 1992. Self-monitoring as an appraisal system for agricultural extension staff. In *Proceedings, Symposium for Research in Agricultural and Extension Education*, Columbus, OH.

Etling, A., and Maloney, T. 1994. *Needs assessment for extension agents and other nonformal educators*. University Park, PA: Pennsylvania State University, College of Agricultural Sciences, Cooperative Extension.

Krueger, R. A. 1994. *Focus groups*. 2nd ed. Thousand Oaks, CA: Sage Publications.

Moore, C. M. 1987. *Group techniques for idea building*. Newbury Park, CA: Sage Publications.

Murphy, J., and Marchant, T. J. 1988. *Monitoring and evaluation in Extension agencies*. Washington, D.C.: The World Bank.

Norland, E. V. Winter, 1991. Handling survey data. *Journal of Extension* 29(4):-37–38.

Patton, M. Q. 1987. *Utilization-focused evaluation*. Beverly Hills, CA: Sage Publications.

Salant, P., and Dillman, D. A. 1994. *How to conduct your own survey*. New York, NY: John Wiley & Sons.

Sanders, J. R. ed. 1994. *The program evaluation standards: How to assess evaluations of educational programs*. 2nd ed. Thousand Oaks, CA: Sage Publications.

Sawer, B. J. 1992. *Evaluating for accountability: A practical guide for the inexperienced evaluator*. Corvallis, OR: Oregon State Cooperative Extension.

Scriven, M. 1993. *Hard-won lessons in program evaluation*. New Directions for Program Evaluation, #58. San Francisco, CA: Jossey-Bass.

Spiegel, M. R., and Leeds, C. August 1992. *The answers to program evaluation: A workbook*. Columbus, OH: The Ohio State University, Cooperative Extension Service.

Smith, M. F. 1989. *Evaluability assessment*. Norwell, MA: Kluwer Academic Publishers.

Treasury Department. November, 1993. *Performance measurement guide*. Washington, D.C.: Financial Management Service, Department of the Treasury.

CHAPTER

> ***"Can there be a humane society without volunteers? Can there be a democratic society without voluntary action? Can there be a free society without voluntarism? I think not."***
>
> —*Leo Perlis*

Management of Volunteer Programs

Key Concepts:

- Why people volunteer
- Volunteer program models
- Developing a volunteer management program
- Volunteer roles
- Future trends and issues

VOLUNTEERS IN EXTENSION

The concept of volunteers is not new. Since the beginning of time, individuals have shared their time and talents for the betterment of community. Volunteering is a practice that is deeply rooted in this nation's history, from clearing the land to organizing the militia and establishing governance. As the country has continued to grow, the volunteer movement has also continued to grow and expand to address ever changing needs.

Volunteers comprise one of the largest groups of untapped resources. Over half of all Americans volunteer in some capacity. According to a national survey conducted by the Gallup Organization (1992), 51 percent of Americans volunteer an average of 4.2 hours per week. The same survey found that 94.2 million adults volunteered a total of 20.5 billion hours in 1991. Those figures had an estimated dollar value (based on an average nonagricultural wage, plus benefits) of $176 billion. Additionally, 61 percent of American teenagers twelve to seventeen years of age were found to volunteer an average of 3.2 hours per week.

Vineyard (1993, p. 3) describes the stereotypical twentieth-century volunteer as middle aged, white, female, middle to high income, compliant, married with children, and not working outside of the home. But times have changed and so has the modern American volunteer. Today's volunteer is more likely to be working at least one job, balancing career and family, have minimal free time, be surrounded with high stress, and carefully evaluate any volunteer assignment before making a commitment. The modern volunteer is equally likely to be male or female, white or nonwhite, single or married, and from any income level.

The 1992 Gallup Organization national survey identifies the following trends in volunteerism:

- African Americans are volunteering in greater numbers than before.
- Although minority groups are asked to volunteer less often, when asked, they volunteer at a rate higher than the average population.
- The majority of volunteers are working people.
- Single persons are among the groups showing significant increases in percentages of volunteers.
- Volunteers give more money to charity than nonvolunteers.
- One-fourth of all volunteers (25.2 million adults) volunteer five or more hours per week.
- Those who identify themselves as religious and who attend religious services regularly are the most generous with their volunteer time.

Just as volunteers are an essential component of the American society, they are the lifeblood of Extension. As part of the land-grant university system, the Cooperative Extension Service was founded on the philosophy of taking the university to the people by providing research-based knowledge to address local concerns and issues. This makes every resident of the United States a potential client. There will never be enough salaried staff to give adequate attention to every individual who comprises this target audience. From its beginning, volunteers have been active partners in the daily operations of Extension. A national study conducted by the University of Wisconsin (Steele 1986) found that one in every eighty Americans (2.9 million) has served as a volunteer in some capacity with the Cooperative Extension Service. Collectively these volunteers invested fifty-one days for every single day an Extension professional invested in working with volunteers. According to the same study, if a dollar value were placed on volunteer time, it would be five times more than the entire Cooperative Extension Service budget, and if communities had to pay for these services provided by Extension volunteers, they would have spent more than $4.5 billion annually. Figure 9–1 illustrates the dollar value by clusters of tasks completed by Extension volunteers. Although the greatest involvement of volunteers is with the 4-H Youth Development program, volunteers are ac-

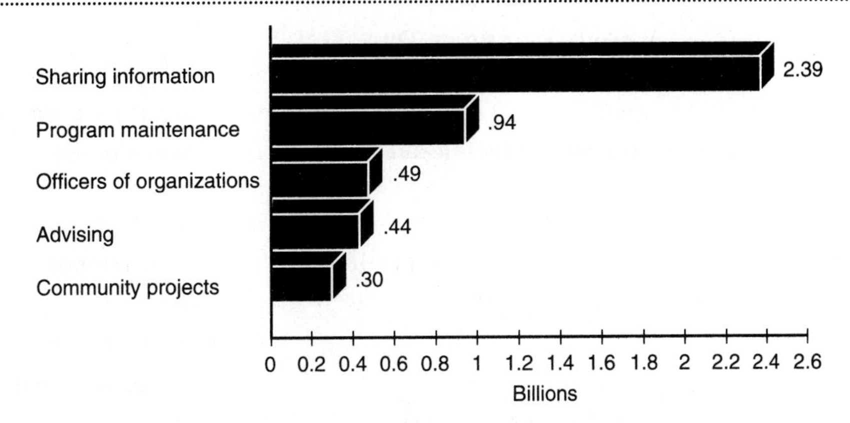

Figure 9–1 ***Dollar Value of Clusters of Volunteer Tasks Volunteers' View of Value. 1985.*** **Partners in action: community volunteers.** ***Cooperative Extension Agents National Projections. University of Wisconsin, Madison.***

tively engaged in programs and activities affiliated with all Extension program areas. Volunteers and Extension faculty team together to make a winning combination.

Why Do People Volunteer?

The mystery of human behavior—attempting to understand why people do the things they do—has perplexed people for centuries. What is it that makes people "tick?" Why is it that, given specific situations or conditions, some people will choose to participate and others will not? People choose to volunteer for many reasons whether it be for the genuine satisfaction and pleasure of giving or whether it be for personal gain and recognition. Buford (1988, p. 145) says human motives "are based on a variety of drives, desires, wishes, and similar forces often called needs." Recognizing the needs individuals bring to volunteer situations can help the program manager match the jobs that need to be done with the interests and talents of the volunteer.

The 1992 national survey by the Gallup Organization asked individuals to identify motivations and positive experiences that caused them to volunteer and/or increase giving. The survey revealed that:

- 69 percent had volunteered before
- 67 percent wanted to make a significant change in society
- 66 percent had belonged to a youth group
- 67 percent saw someone they admired helping another
- 61 percent saw someone in their family helping others
- 60 percent were helped by others in the past

A collective review of the literature (Anderson and Lauderdale 1986), provides a more exhaustive list of why people volunteer.

People Volunteer To:

belong	have a part in problem solving
give	test out career possibilities
complete educational requirements	receive evaluated work experience
be recognized	be creative
receive professional training	make new friends
acquire new skills	use existing skills
be part of a team	break from boredom and monotony
have fun	develop personal leadership ability
develop new interests	

Regardless of the motivation, the primary reason people follow through is because they were asked! A vast wealth of talent and resources is available when the needs are clearly identified and people are asked to help. Everyone wins when an effective volunteer program is in place. Volunteering has benefits for the program and its clientele, for the community, and for the volunteers themselves. An obvious benefit is economic savings due to enhanced resources. Volunteers are essential partners who can extend the roles of paid staff to reach larger audiences, conduct additional programming, and provide creativity and diversity to the work environment. Communities benefit from the bonding that occurs and linkages that develop when citizens join together in community problem solving. Finally, as already discussed, the volunteers themselves are direct beneficiaries as a result of personal growth and development, increased workplace skills, and a sense of fulfillment.

VOLUNTEER PROGRAM MANAGEMENT

An Extension professional's job is varied and complex. Regardless of technical areas of expertise, an essential competency for all Extension professionals is management and direction of volunteer programs (Hahn 1979). In 1983, the average Extension agent spent about a third of his or her work time with volunteers. Establishing and maintaining an effective volunteer management program requires a planned, ongoing approach. Several volunteer management models have been established, but two models that have been used frequently in Extension are Boyce's ISOTURE model (1971) and Penrod's L-O-O-P model (1991). The ISOTURE model is comprised of seven separate but interrelated volunteer functions: identification, selection, orientation, training, utilization, recognition, and evaluation, see figure 9–2.

First introduced in 1991, the L-O-O-P model is an acronym for **locating** (selection and recruitment), **orienting** (informal and formal), **operating** (education and accomplishment), and **perpetuating** (evaluation and recognition), see figure 9–3. Each of four functions represents a different

Establishing and maintaining an effective volunteer organization requires a planned, ongoing approach that can be accomplished using the following steps:

I	=	**I**dentification
S	=	**S**election
O	=	**O**rientation
T	=	**T**raining
U	=	**U**tilization
R	=	**R**ecognition
E	=	**E**valuation

Figure 9–2 ISOTURE: A Volunteer Development Model

phase in a sequence of the management process. Arrows between the functions indicate that the activities are intended to blend together and are not independent of one another.

These two models and others found in the literature all identify similar functions or phases to follow. Following a logical sequence enhances the probability of an effective volunteer program. Coordinating staff (both paid and volunteer) is equivalent to the role of a mechanic. A program that runs smoothly is like a well-oiled machine. Each separate piece or part has a specific function. When each separate piece is functioning as it should, the machine is able to carry out its operations. However, each separate piece influences all others. When one is not functioning, the entire machine fails to operate at maximum capacity or at worst breaks down completely. Using the ISOTURE model as a guide, the following sections will address the major functions of managing an effective volunteer program.

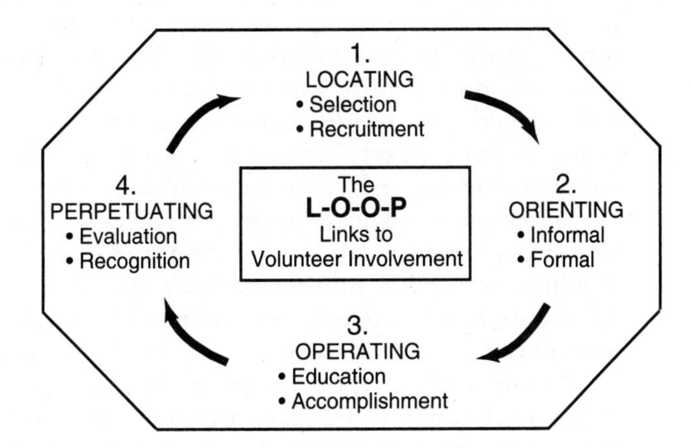

Figure 9–3 The L-O-O-P Model. Penrod, Kathryn. 1991. **Journal of Extension.**

DEVELOPING A VOLUNTEER MANAGEMENT PROGRAM

Identification involves identifying volunteer opportunities within the organization and developing appropriate written job descriptions for them. Developing a volunteer management program requires planning and organization. Managers may have a strong urge to begin with recruiting volunteers. They should resist and begin the process with a thorough look within the program or organization. Specifically, they should understand the needs of the organization or program, how they can best be served, and where volunteers fit in this scenario. Once it has been determined what volunteer jobs are needed, each job should be analyzed to know what kind of volunteer to look for. Once types of roles or positions have been identified, specific job descriptions should be developed. Job descriptions are a critical communication tool. Their main purpose is to clearly identify the expectations of the job and describe the tasks to be performed. Although job descriptions will vary in format and length, they should contain the following basic information:

1. Title—identify job with a specific title
2. General Description—sponsoring program, overall responsibilities of volunteer
3. Skills—required (or desired) of person(s) doing the job
4. Specific Responsibilities—concise list of what is to be done
5. Expected Results—impact due to job being done
6. Resources—human and material resources available
7. Supervision—identify supervisor by name and title
8. Location—where the volunteering will take place
9. Time Involved—hours, days, weeks or months of the year

Although developing job descriptions for each position that needs to be filled may seem like an overwhelming task, the benefits of such an effort are great. Not only does it add a business-like dimension to the situation, but most people want to know exactly what is expected of them before making a commitment. Job descriptions are also useful documentation in the unlikely situation of needing to fire a volunteer. A sample volunteer agreement can be found in figure 9–4.

Selection involves choosing the best-qualified individual for the volunteer opportunity. Although in the true sense "volunteer" means that individuals self-select, many volunteer roles are filled through recruitment efforts. Research suggests that the most successful recruitment efforts are a result of personal contact—people will volunteer when they are asked.

A good job description is an excellent recruitment tool. When specific needs are known, individuals who have the necessary qualifications can be appropriately recruited. For example, if you need someone with accounting experience, check the business community. If you need a chap-

TITLE: 4-H Contact Helper—LABO Exchange date: revisions:

PURPOSE: Contact person for LABO Exchange in Adams County

VOLUNTEER:

VOLUNTEER RESPONSIBILITIES:

- Keep in touch with LABO program coordinator re: program plans.
- Promote LABO program.
- Disperse applications to potential host families.
- Review applications.
- Enroll host families in International Intrigue Project.
- Inform families of selection and assignments.
- Keep in contact with families during hosting period.

VOLUNTEER QUALIFICATIONS:

- Interested in people.
- Experience with LABO.
- Communications Skills.

ADVISOR: LABO Program coordinator

ADVISOR RESPONSIBILITIES:

- Keep contact helpers informed.
- Coordinate with Japanese staff re: plans.
- Assign exchangees to host families.
- Arrange transportation and group programs for LABO delegation.
- Handle any delegate and hosting problems referred by contact helpers.

TRAINING AND/OR RESOURCES PROVIDED:

- International Intrigue Project training.
- LABO Program Update (ongoing).
- Consultation as needed with LABO Program coordinator.
- Access to Extension Office for telephone and mail needs.

TIME ESTIMATE: 3 hours/week for one year.

OTHER NOTATIONS: Time may not be evenly spread since certain deadlines will require more effort at times.

Figure 9–4 Sample Volunteer Agreement

eron, ask a parent of one of the participants. According to Steele (1989), there are three major sources of potential volunteers: people who are directly involved or know someone who is, people with special talents they are willing to share, and people who are generally willing to help regardless of the program or cause. Personal contacts, flyers, notices, recommendations, and newspaper advertisements are all proven methods for recruiting volunteer staff. An important rule to follow, however, is not to recruit until you know what you are doing and can communicate to the potential volunteer what he or she will be doing.

More than ever, there is stiff competition to attract the best volunteers. Competition lies with other agency programs, changing life styles, and busy work and family schedules. Recruitment involves trying to remove people's reasons to say no, without twisting their arms into volunteering. It is important to help people understand not only what they will do and how long it will take, but also the purpose of the activity or event and who will benefit from it. Be honest with people when recruiting. Do not lie about or minimize the work or time needed and never use guilt to get people to commit.

Selection is a two way process. Not only is the professional staff attempting to fill a specific role, but the volunteer also is attempting to find the position that will best satisfy personal interests and motives. Satisfaction and productivity will increase when a match occurs that meets everyone's needs. Determining a good match requires obtaining information. Many volunteer applicants are required to complete a formal job application, including education and work experience; prior volunteer experience (including where and the type of work); special skills, interests, and hobbies; motivation for volunteering; ability to provide own transportation if necessary; and information regarding availability for volunteer work. Interviews with applicants provide a more personal means of gathering information. The interview should build on the information obtained from the application form. Recently many programs have begun utilizing strict selection processes in order to minimize any potential threats to clients. Negative publicity regarding volunteer staff involved in criminal acts, abusing or harassing clientele, and spreading contagious diseases has raised significant concern that prompted action. Specific programs (commonly health and youth programs) require potential volunteers to submit to medical and criminal screening and to sign a contract of agreement, see figure 9–5.

The types of volunteer roles an individual may assume in Extension work are many and varied. Although the largest portion of volunteers in Extension assist with the 4-H Youth Development program, volunteers are active in every program area. Figure 9–6 identifies seven extension volunteer programs or roles that individuals select. These programs account for about half of all work with volunteers. Agents work with other volunteers on short-term projects and activities, as well as with those on advisory committees, boards, and councils. They also work with many volunteers

This Standard of Behavior is contractual agreement accepted by volunteers who commit to the 4-H program. The Standards shall guide their behavior during their involvement in Ohio 4-H. Just as it is a privilege for The Ohio Cooperative Extension Service to work with individuals who volunteer their time and energies to Ohio 4-H, a volunteer's involvement in Ohio 4-H is a privilege and a responsibility, not a right.

The Ohio 4-H program provides quality educational programs accessible to all Ohio youth. The primary purpose of this Standard of Behavior is to insure the safety and well-being of all 4-H participants (i.e., members, their parents and families, professionals, and volunteers).

Ohio 4-H volunteers are expected to function within the guidelines of The Ohio Cooperative Extension Service and the Ohio 4-H program. Ohio 4-H volunteers shall be individuals of personal integrity.

Ohio 4-H volunteers will:

- Uphold volunteerism as an effective way to meet the needs of youth and adults.
- Uphold an individual's right to dignity, self-development, and self-direction.
- Accept supervision and support from professional Extension staff while involved in the program.
- Accept the responsibility to represent their individual county 4-H program and the Ohio 4-H program with dignity and pride by being positive mentors for the youth with whom they work.
- Conduct themselves in a courteous and respectful manner, exhibit good sportsmanship, and provide positive role models for all youth.
- Respect, adhere to, and enforce the rules, policies, and guidelines established by their individual county 4-H program, the Ohio 4-H program, and The Ohio Cooperative Extension Service.
- Recognize that (1) abuse of *any* 4-H participant by physical or verbal means, (2) failure to comply with equal opportunity and antidiscrimination laws, or (3) a felonious criminal act *will* be grounds for termination as a 4-H volunteer. Failure to discharge duties in a responsible and timely manner *may* be grounds for termination as a 4-H volunteer.
- Report immediately any threats to the vounteer's emotional or physical well-being to the county 4-H professional.
- Accept the responsibility to promote and support 4-H in order to develop an effective county, state, and national program.
- Operate machinery, vehicles, and other equipment in a responsible manner.

I have read and understand the Standards of Behavior outlined above. I understand and agree that any action on my part that contradicts any portion of these standards is grounds for the suspension and/or termination of my volunteer status with the Ohio 4-H Program.

Signature of 4-H volunteer ______ Date ______ Signature of Extension Professional ______ Date ______

Figure 9–5 Ohio 4-H Volunteer Standards of Behavior

from other agencies and organizations as well as with the thousands of people who just "lend a hand."

Orientation familiarizes the volunteer with both the total organization and with the specific job responsibility. Every volunteer should be involved in some form of orientation. Orientation is an essential element in the volunteer management process because it is the initial training session for all new volunteers. An important fact to note is that willingness to participate is completely separate from the ability to complete the task. It is entirely

Title of Program	% of U.S. Counties carrying	Approximate number of volunteers	Description of the Program
Local 4-H Leaders	96%	524,000	Work with youth groups on a continuing basis on organizational procedures, teaching projects, and guiding participation in events
Middle Management 4-H	44%	35,800	Assist the Extension staff in training and counseling local 4-H club leaders and conduct countywide events.
Extension Homemakers	79%	269,000	Teach family-related information to local Extension Homemaker clubs; work on community projects
Master Gardeners*	26%	15,100	Answer individual questions, serve at plant diagnosis clinics, teach groups, work with community horticultural projects and events.
Master Home Economics Volunteers*	21%	9,700	Teach family-related subject matter in the same way as Master Gardeners. Title varies with subject matter. For example, Master Food Preserver, Master Financial Counselor.
Ag Cooperators	96%	16,500	Assist with events, teach groups, provide funds and facilities, work with test plots and agriculture-related projects.
Resource Volunteers	66%	11,400	Spearhead community projects including new facilities, surveys, policies.

*Master Volunteers give a specified number of volunteer hours in return for a specified number of hours of training. Some programs require Master Volunteers to meet certification requirements.

Figure 9–6 ***Volunteer Programs and Roles in Cooperative Extension. Volunteers' View of Value. 1985.* Partners in action: community volunteers. *Cooperative Extension Agents National Projections. University of Wisconsin-Madison.***

possible for someone to be willing to assist but not be trained in the specific skills necessary to complete the effort. At the same time, there may be many who possess the desired skills, but for whatever reason are unwilling to share their talents. Establishing a positive environment (both physical and psychological) during orientation is critical for comfort and success in the organization. Depending on the situation, orientation of new volunteers may be accomplished at a general session or meeting, on a one-to-one basis, or through self-directed inquiry from the individual (questions, reading policy manuals, etc.). Whether it is completed in one session or over a period of time, specific topics should be addressed. Those topics include:

- Description and history of program (why are we here?)
- Organization chart and introduction to key staff
- Supervision system (who reports to whom, process for handling complaints/concerns)

- Review of policies and procedures (including liability and insurance)
- Facilities (parking, meals, rest rooms, phones, etc.)
- Volunteer benefits (including recognition)
- Record-keeping requirements
- Emergency procedures
- Expectations and responsibilities of individual job
- Procedures for schedule changes or absences

Training assists volunteer staff in developing additional knowledge, attitudes, and skills that will help them do their jobs better and more responsibly. Training will vary greatly depending upon the level of skill the volunteer brings to the situation. Wilson (1988) describes three levels of training programs. The first is prejob training that is designed to prepare an individual for a job before he or she begins. Master volunteers for programs such as gardening and food preservation participate in hours of training before they are permitted to answer clientele questions. On-the-job training allows an individual to enhance knowledge or improve skills while actually on the job. Continuing education is a lifelong process. Opportunities for personal growth should be available through individual inquiry and through opportunities provided by the organization. Regardless of the approach, training should be relevant to the situation, appropriate to the level of the learner, and interesting.

Utilization involves utilizing volunteers' knowledge, skills, and attitudes to contribute to the success and growth of the organization. Utilization is closely tied to the components of good supervision. The good supervisor is one who is able to get people to do an effective job. The measure of a supervisor's success is determined by how well his or her people do their share of the work and how high their morale is while they do it. Safrit (1992) defines utilization as the "putting people to work" phase. Effective utilization involves (1) placement—the right person in the right job with the right preparation, (2) delegation—provision of authority and guidance to do the work, (3) follow-up—regular training or updates to ensure progress, and (4) communication—ongoing, two-way communication between volunteer and staff member.

Is supervision of volunteer staff different from supervising paid staff members? The answer is yes and no. A primary distinction that influences approaches to supervision is the volunteer's motivation for participation. Earlier, this chapter discussed several motivational theories that identified different reasons for participation. Significant differences in attitude, performance, and efficiency can occur when (1) there is a fear of dismissal or reprimand (paid staff); (2) participation is required (class credit or court-ordered community service); (3) a high need exists for recognition as often exemplified through extrinsic measure; (4) social needs are high (desire to meet people); or (5) participation relates to fulfilling some intrinsic needs or interests (satisfaction or fulfillment, making a difference). Understanding what motivates a volunteer can assist the supervisor in not only making a better match to a position, but also in selecting supervision strategies. Those with high social needs should not be placed in isolation, but rather be placed in a position that allows interaction and productivity. Volunteers are most involved because they want to be. If their needs are not being met, they can leave. A supervisor needs to understand the motivation to participate, interact with those supervised, provide constructive feedback and support as opposed to criticism, and include them as part of the team. Respect the value of their contributions and recognize and reward them. Interaction develops trust and respect. It assists in determining how, when, or if to delegate responsibility, and the ability or complexity level of the job to be done. Open communication encourages support, team effort, and minimized conflict.

Recognition of volunteer contributions to the organization is an essential component of volunteer management. Volunteers and their positive contributions are often taken for granted. They and the services they provide often go unrecognized and unthanked. Appreciating the positive contributions that volunteers make and recognizing them for it is important. Showing appreciation is showing respect for and acknowledging the value of volunteers.

Volunteer recognition comes in many different forms. Formal approaches are those that are most tangible: certificates, pins, tokens, plaques, letters of appreciation, media releases, public acknowledgments, and banquets. Informal recognition is often more spontaneous and intangible and may include words of praise or consideration for special opportunities, such as taking on additional responsibilities, participation in special training programs, or being asked to represent the agency or program. The best time for recognition is immediately following the completion of the task or activity. The most effective recognition efforts are ongoing. Most people don't volunteer for the recognition they hope to receive; however, it's a good feeling to know that time and effort are appreciated. Although the reasons people stop volunteering are as many and varied as the reasons they start, a volunteer who is recognized and appreciated for his or her efforts is more likely to stay with the organization.

Evaluation of the individual's performance as a volunteer as well as of the total volunteer development program is the final phase in the ISOTURE model. Similar to other staff members, volunteers should receive feedback regarding the quality of their work, including their existing strengths and areas needing improvement. Only by acknowledging areas for improvement can individual and organizational growth occur. Recognition is not to be confused with evaluation. It is important to recognize participation and efforts, but recognition does not provide a complete assessment of the performance or program. Informal techniques including discussion and short interviews are the most frequent methods used in evaluating Extension volunteers. Formal strategies include self-rated checklists, supervisor appraisals, peer and clientele reviews, and in-depth interviews.

FUTURE TRENDS AND ISSUES

We live in a rapidly changing world. Volunteer programs, like most other programs, reflect the changes and trends of the times. Population shifts, management trends, social issues, and government have all had their impact on the management of volunteer programs.

Diversity

Diversity is the buzz word of the 1990s. Initial perceptions of diversity link it to racial and ethnic background. True diversity, however, recognizes and respects the wide variety of differences individuals and groups possess and encourages harmonious existence by respecting the differences rather than attempting to eliminate or ignore them. Census data confirm that ethnic diversity of the United States is on the rise. This trend is also reflected in volunteer participation. The 1992 Gallup survey identified an increase in volunteerism among black populations. The same study also found that volunteer participation among minority groups overall was lower than among Caucasians, but that when asked directly to volunteer, minority groups had a higher rate of acceptance. A misconception, however, is that diversity relates only to racial or ethnic background. Every person is a unique individual. Diversity extends beyond heritage to include gender, age, religion, geographic region, physical abilities and disabilities, education and economic levels, and lifestyle choices. The stereotypical volunteer of the 1950s and 1960s no longer exists. The volunteer pool available today is more reflective of the population it serves. Another area where these changes are visible relates to age. Greater numbers of youth and retirement-age individuals can be found volunteering their time and service. Volunteer administrators today need not only to be sensitive to the issues of diversity but prepared to expand and integrate their present volunteer pools.

National Service

The Community Service Act of 1990 authorized funding for model programs with the vision of developing a national service initiative. Intending to utilize existing state and local networks, the initiative is to focus on unmet educational, environmental, human, and public safety needs. National service efforts are directed toward all citizens; however, initial efforts have concentrated on youth. National service will have an impact on existing volunteer programs. A significant impact will be an influx of volunteers from all walks of life, backgrounds, and interests that will join the existing corp of volunteers already serving America. Vinyard (1993) says,

> *the call to service has gone out and is being heard, by individuals, by groups, by suddenly-interested parties now eager to put together programs that can meet needs from grassroots to mountaintops. The formalized National Community Service initiatives are simply vehicles in a long and varied line of efforts that bring together the needs of our land and the people willing to serve those needs. The 1990s and turn of the century will be remembered as the Decade of Service.*

Volunteer Liability and Legal Implications

Nationally the number of trained attorneys and the number of legal actions has dramatically increased in recent years. Volunteers have not been immune to lawsuits and as a result some hesitancy exists in both those considering volunteering and those responsible for directing volunteer programs. In almost all instances that involve legal issues, volunteer staff are considered equal to paid staff. Despite the fact that volunteers do not receive pay, they accept responsibility for agreed-upon tasks and, as such, are potentially liable for their actions while fulfilling that job. Additionally, as in any employment situation, agencies that use volunteers are liable for the actions of their volunteers. An increase in legal actions throughout the United States involving both paid and volunteer staff has caused many to be wary. As a result, more and more agencies have adopted many of the volunteer practices identified throughout this chapter.

Detailed job descriptions clearly outline the job responsibilities and qualifications. Interviewing candidates and requiring volunteers to sign a contract of agreement further support the commitment of the agency to place highly qualified personnel. Additional measures commonly used include screening for criminal activity and health-related concerns. Many of these practices are costly and time consuming to implement and coordinate, and they do not guarantee that legal action will not occur. They do, however, communicate a message of concern and commitment to the program, its clientele, and the potential volunteer. In most cases, the practices serve as a deterrent to individuals whose motives are not pure. Many agencies and programs that involve volunteer staff purchase liability insurance. Accurate record keeping by the agency is necessary to identify the volunteer's affiliation with the agency.

Volunteers who serve on local boards, councils, and advisory committees have some unique distinctions. Most members of boards and committees have a rotating short-term assignment. Regardless, boards, councils, and advisory committees function to determine policy and to advise programs and procedures that can have lasting implications. Liability for actions can extend beyond the board as an entity to its individual members. Agencies and programs that utilize boards or advisory committees should determine what legal representation is available. Individuals asked to serve on a board should determine what their personal liability is. Agencies, managers and volunteers should not be afraid to become involved. Utilizing sound management practices and being aware of rights and responsibilities provides a solid foundation for an effective volunteer management program.

SUMMARY

Many different volunteer development models exist. A successful volunteer program requires time, commitment, and energy. It does not just happen. It requires careful thought, planning, and organization. Volunteers are essential for an effective Extension organization and for the welfare of the United States. Volunteers are part of our rich heritage. More importantly, they are a critical force for the future.

DISCUSSION QUESTIONS

1. What are some of the types of roles volunteers assume in your community? What are some specific roles they assume within the Cooperative Extension Service?
2. What are the advantages of volunteering to (1) the organization and (2) to the volunteers themselves? Are there any disadvantages?
3. What are some alternatives to "firing" a volunteer?
4. Should volunteers be required to submit to screening procedures?
5. Are there certain roles within Extension that volunteers should not assume? Why or why not?
6. With people being asked to volunteer for more programs, what are some unique ways to recruit volunteers?
7. Volunteer retention is as important as recruitment. What conditions need to be present to assure volunteers continue with your program?
8. Should volunteers be subject to a formal performance appraisal?

REFERENCES

Anderson, S., and Lauderdale, M. 1986. *Developing and managing volunteer programs.* Springfield, IL: Charles C. Thomas.

Blanchard, K., and Johnson, S. 1985. *The one minute manager.* New York, NY: Berkley Books.

Boyce, M. 1971. *A systems approach to leadership development.* Washington, D.C.: United States Department of Agriculture, Cooperative Extension Service - 4-H.

Buford, J.A., and Bededian, A. G. 1988. *Management in Extension.* 2nd ed. Auburn, AL: Auburn University: Alabama Cooperative Extension Service.

Gallup Organization. 1992. *Giving and volunteering in the United States, 1992.* Survey. Washington, D.C.: Gallup Organization Independent Sector.

Hahn, C. P. May 1979. *Development of performance evaluation and selection procedures for the Cooperative Extension Services: summary report.* Prepared for the U.S. Department of Agriculture. Contract Number 12-05-300-372. Washington, D.C.: American Institutes for Research.

McClellend, D. C. 1987. *Human motivation.* Cambridge, MA: Cambridge University Press.

Penrod, K. M. Winter 1991. Leadership involving volunteers. *Journal of Extension* Vol. 28. 9–11.

Perlis, Leo. September/October 1974. p. 1 *Voluntary action news.*

Safrit, R. D., and Smith, W. 1992. *Building leadership and skills together.* Columbus, OH: The Ohio State University Extension, 4-H.

Steele, S. M. 1986. *Partners in action: a national study of volunteerism in Extension.* Cooperative Extension System National Accountability Study. Madison, WI: University of Wisconsin.

Vineyard, S. 1986. *101 ideas for volunteer programs.* Downers Grove, IL: Heritage Arts Publishing.

_______ V. 1993. *Megatrends & volunteerism.* Downers Grove, IL: Heritage Arts Publishing.

Wilson, M. 1988. *The effective management of volunteer programs.* Boulder, CO: Volunteer Management Associates.

CHAPTER

"The measure of the value of a nation to the world is neither the bushel nor the barrel, but the mind."

—*Sir William Osler*

International Extension

KEY CONCEPTS

- Importance of international extension
- Characteristics of various extension models
- Constraints to extension worldwide

INTRODUCTION

The goal of the 1914 Smith-Lever Act, ". . . to aid in diffusing among the people . . . useful and practical information . . . [through] instruction and practical demonstrations," has helped improve educational levels and the quality of life around the world. There are extension systems in 184 countries (Swanson 1990), and they are based upon many of the same concepts and principles as the United States model.

In addition to similarities, there are differences in extension systems from country to country. The United States was not the first to implement an extension system, nor is the land-grant university model of Cooperative Extension the only one in use internationally. This chapter will group the different ways to organize extension work into four models, list the characteristics of each, and identify some of the constraints faced by extension worldwide. But first, before discussing the models and constraints, we need to understand why it is important to study extension education in countries outside the United States.

REASONS TO STUDY EXTENSION IN OTHER COUNTRIES

There are at least three reasons why Extension educators should be interested in extension internationally:

- opportunities for careers and experiences abroad
- gaining ideas from other systems
- educating clients on international concerns

Opportunities for Careers and Experiences Abroad

Studying international extension is important because of the career opportunities available for those with experience in U.S. extension work. Persons with extension backgrounds, who know how to organize commodity groups, and who know about the inputs required can help expand development overseas. How U.S. extension has succeeded is a story that others are eager to hear. A word of caution, however: an educational program that works well in one culture will not necessarily be successful in different circumstances.

Exchanges and short-term experiences for U.S. extension personnel and clients help to promote awareness of other cultures and international understanding. Exchanges of 4-H young people under the International Four-H Youth Exchange (IFYE) program have been in place since 1948 (Rasmussen 1989). Also, state and local U.S. extension personnel assist in training and field experiences for international visitors and help to place exchangees with families.

Gaining Ideas from Other Extension Systems

Another reason for studying about extension internationally is that knowledge about how other systems operate can help improve one's own. Like any good idea, extension education has many forms and adaptations.

Internationalizing Extension Education Programs

The third reason for internationalizing extension is to educate clientele about the interdependence of world economies, the widespread effects of environmental damage, and the need to preserve the gene pool of agricultural products. Food security, economic well-being, and world peace are closely interrelated (Wisner and Wang 1990).

Production is not the only factor in providing enough for everyone; also to be considered are distribution, storage, and population growth. Malthus, a British economist, predicted in 1787 that the number of people would increase faster than the amount of food. This has not happened, but the current anticipation is for relatively rapid population growth rates in many major rice-consuming nations (Wisner and Wang 1990). The world's need for food to feed the population and diminish hunger and malnutrition is expected to be 50 percent more in the year 2000 than it was in 1979 (Food and Agriculture Organization 1979).

How countries conduct their agriculture has an effect not only on the world's markets but also on the world's environment. Alterations in the environment, whether by human activity or climatic changes, are global changes with implications for the food supply and the livelihood of farmers, foresters, and ranchers. Desertification, destruction of the major rainforests, and range overexploitation are examples of activities that threaten environmental stability (Cooperative States Research Service 1993).

Other countries are important to the protection of the gene pool of agricultural plants and animals. Protecting the diversity of the gene pool provides insurance against disease, pests, and extremes of weather that might decimate highly inbred crop varieties. The monocultural agriculture practiced in the United States (huge fields of a single cultivar of a single crop) needs to be balanced by the diversity found elsewhere.

ORIGINS OF EXTENSION WORLDWIDE

The United States was not the first to have a formally organized Extension Service. That honor goes to Japan, which began its service in 1893, see figure 10-1. The dates in figure 10-1 are official starting dates of a few of the early systems; in all cases extension activities on a limited scale preceded all of the national establishment dates.

Approximately 54 agricultural extension organizations were established between 1900 and 1959 (Swanson 1990). From 1960 to 1990, an additional 130 extension organizations came into existence. There were several reasons why many countries established extension systems in the second half of the twentieth century:

1. Many former British colonies achieved independence, and newly independent countries reorganized existing agricultural ministries to include an extension unit.
2. After World War II ended in 1945, the United States and others began massive international aid efforts. Technical assistance emphasized agricultural growth and helped establish extension services.
3. Increased trade and economic expansion were available to pay for extension.

COUNTRY	YEAR
Japan	1893
United States	1914
United Kingdom	1946
India	1952
Egypt	1953
Nigeria	1954
Taiwan	1955
Brazil	1956

Figure 10-1 ***Establishment of National Extension Services in Selected Countries. George Axinn and Sudhaker Thorat. 1972.* Modernizing World Agriculture. *New York: Praeger Publisher.***

4. A backlog of agricultural research information was there to be disseminated.
5. Mass-media communication and modern transportation made it possible to provide extension services more efficiently.

COMPARISON OF U.S. EXTENSION AND EXTENSION IN OTHER COUNTRIES

Extension approaches can be divided into four general categories:

- U.S. Model
- Developing Country Model
- Chinese Model
- European Model

All models of extension have some common characteristics: a sponsoring organization, transfer of information to audiences, and promotion of change. They all have personnel responsible for transferring information and promoting change. Some of the challenges to extension organizations are the same everywhere: budget constraints, environmental concerns, rural-to-urban migration, need for linkages between research and Extension, governmental policies that conflict with individual welfare, and communities and families with economic and social problems.

In most countries, the sponsoring organization for extension is the Ministry of Agriculture, which would be comparable to the U.S. Department of Agriculture. However sponsors may also include governmental organizations at different levels, educational institutions, and private as well as public organizations. Figure 10–2 illustrates some of the differences between the U.S. model and the developing country model.

Philosophical Differences

Some of the differences in extension systems internationally are due to differences in philosophy. One view is that nations with a food shortage should expand their productive capacities. This view believes that it is important for each political entity to be self-sufficient in food. Another view is that agricultural commodities should be produced in the most favorable regions and transported to other countries, regardless of whose farmers are hurt in the process. Yet another view, held by some policy-makers, is that U.S. aid to hungry countries should be based on what is currently in surplus, those commodities for which the government is paying price supports. Humanitarians want to ensure that no one starves, but economists argue that giving food to other countries destroys their prices and disrupts their supply systems. Some governments are interested mainly in extension efforts that will increase their agricultural exports and consequently their balance of trade. Others are more concerned about resettling their small farmers and trying to reverse urban migration. Each of these philo-

DEVELOPING-COUNTRY EXTENSION	U.S. COOPERATIVE EXTENSION
ORGANIZATION	
Ministry of Agriculture Hierarchical organization	University-based Flat organization
Inputs of credit, supplies, transportation, chemicals in short supply	Private sector inputs of seed, fertilizer, or credit
Ill-trained staff	Highly educated staff
Outside funding	Funded by local, state, national government
PROGRAM CONTENT	
Emphasis on agriculture	Includes agriculture, home economics, youth, and community development
Education, regulation, and provision of inputs, such as seed and fertilizer	Education
Often weak linkage to research	Strong research base
Emphasis on productivity	Emphasis on profitability
National production goals	Needs of individuals and communities
Top-down approach	Locally planned programs
AUDIENCES	
Farmers and rural areas	Includes rural, suburban, and urban programming
Male audiences (Rural youth and women are only 3 percent worldwide)	Youth and adults, male and female
Large farmers	Middle income
More than 50 percent of population engaged in farming	Less than 2 percent of population engaged in farming
Uneducated clientele	Educated clientele

Figure 10–2 ***Extension in Developing Countries Compared with U.S. Cooperative Extension. Julia Gamon, Department of Agricultural Education & Studies, Iowa State University, Ames, IA.***

sophical views has proponents who extol its strengths. Which view is currently in vogue usually depends upon which political party is in power.

When international students were asked their views on Extension's educational philosophy and mission, they wanted their home countries to focus more on empowering people to solve local problems. In other words, the international students preferred a "bottom-up" approach arising from the viewpoint of local communities rather than a "top-down" approach from government agencies (Mohamed, Gamon, and Trede 1995). However, economically underdeveloped countries have a large percentage of their population engaged in farming, see figure 10–3. Where systems of education, technology, transportation, and communication are limited, a central extension system may be needed to ensure that farmers have credit, supplies of seed and fertilizer, and access to markets.

COUNTRY	PERCENT
Kenya	75
Bangladesh	65
China	60
India	65
Nigeria	54
Egypt	34
South Korea	21
Brazil	31
Greece	23
Saudi Arabia	16
Japan	7
Australia	6
Netherlands	4
United States	2

Figure 10–3 Percentage of Population in Production Agriculture. **The World Factbook.** *1995. Washington, D.C.: Central Intelligence Agency.*

DONOR AGENCIES

Because of limited resources in developing countries, which include most of Africa, Asia, and Latin America, support for extension services has come from donor agencies, often with United States dollars, both public and private. The main agencies of the United Nations that support extension programs are the following:

- Food and Agriculture Organization (FAO)
- International Fund for Agricultural Development (IFAD)
- The World Bank

The United States provides extension assistance through its Agency for International Development and through private contributions to nongovernmental organizations, such as Partners for International Education and Training.

Food and Agriculture Organization

The initiatives of the Food and Agriculture Organization of the United Nations (FAO) stress country-specific design, farmer participation, and strong research-extension links. FAO has a strong interest in reaching women farmers and encourages cooperative funding of extension efforts by central, regional, and local contributions (The World Bank 1990). FAO is willing to use a variety of models and provides technical assistance to more than eighty countries.

International Fund for Agricultural Development

The U.N. International Fund for Agricultural Development (IFAD) is dedicated to agricultural research and small farmers. It emphasizes relatively high agent-farmer ratios and extension education for women, and includes long-term national commitment to financing the system.

The World Bank

The World Bank is the largest source of funds for agricultural extension in developing countries, although extension takes only 1 percent of The World Bank's budget (Hayward 1990). The World Bank's support of agricultural extension began in July 1964 with credit to Kenya to pay agricultural instructors to replace departing expatriates.

U.S. Agency for International Development

USAID, the acronym for United States Agency for International Development, is concerned about the incomes of small farmers and emphasizes the need for close links between research and extension and the need to tailor support to local conditions.

Nongovernmental Organizations

Nongovernmental organizations (NGOs) include a variety of U.S. religious and humanitarian groups that are interested in bettering the condition of a people through assisting its agricultural sector. There are many of these; for example, more than 400 NGOs are in Bangladesh, and 90 percent of these are agricultural-development programs (Amanor and Farrington 1991). Nongovernmental donors include foundations, religious groups, and consortiums. Examples are the Kellogg Foundation, Partners for International Education and Training (PIET), and the Academy for Educational Development (AED). Their developmental efforts exist parallel to and in cooperation with governmental extension efforts in Third World countries. Many effective NGOs use networks of grassroots organizations or form new local groups such as agricultural cooperatives, village committees, or irrigation associations (Mattocks and Steele 1994). They are sometimes better able than extension to reach resource-poor farmers and their families. An NGO worker in Zimbabwe or Malawi, for example, might perform activities of a participatory nature, see figure 10-4. Rather than emphasizing technology transfer, the worker (change agent) serves as facilitator and brings the farmer into the process as a partner, not a recipient.

EXTENSION APPROACHES IN DEVELOPING COUNTRIES

All of Africa plus much of Asia and Latin America consists of developing countries, often called the Third World. Within each country, many different approaches have been in use. Axinn (1988) found that there was no "best approach." The main approaches are the following:

- Training and Visit
- Project Approach

ROLE/CONCEPT	TECHNOLOGY TRANSFER	PARTICIPATORY-PROACTIVE
Farmer	Recipient	Partner
Change Agent	Conduit	Facilitator
Method	Reactive/Selling	Interactive/Brokering
Researcher	On-station (unless adoption problem)	On-farm (unless responding to farmer as co-learner)
Management	Top thinks; Local acts	Thinking and acting at all levels
Vision	Top-down	Shared
Approach	Prescriptive	Participation
Evaluation	Summative cost reduction	Formative farmer satisfaction

Figure 10–4 Development Paradigms. David Mattocks and Roger Steele, Winrock International Institute for Agricultural Development, Morrilton, AR.

- Farming Systems Research and Extension
- Participatory

The ministry of agriculture controls extension in almost all countries outside the United States. Contrary to U.S. Extension's land-grant university model of cooperation among federal agency, state university, and local government, extension organizations in Third World countries are organized at all levels through their ministries of agriculture. The ministry of agriculture model is a vertical, hierarchical, model that is compatible with the Training and Visit System of extension, see figure 10–5.

Training and Visit System (T & V)

The Training and Visit System is a wide-spread extension system developed and funded by The World Bank, since the late 1970s. It has been adopted by more than 40 percent of developing countries (Hayward 1990). It is a centrally controlled operation in which Village Extension Workers visit individuals or groups of farmers regularly once every two weeks, see figure 10–6. The extension workers themselves receive training at two-week intervals from a team of Subject Matter Specialists. T & V is an expensive, personnel-intensive, top-down approach.

Advantages of the T & V System are the concentration of efforts on key improvements in major crops and regular contacts. The system has been particularly successful in India and Kenya (Rajasakeran and Martin 1990). Problems with T & V occur when information is too late, distorted, and irrelevant (Diamond 1994). For example, weather conditions may dictate that farmers start planting early, but the relevant T & V training arrives at normal planting time.

Diamond's study of the modification of the T & V System in Swaziland documented the importance of tailoring any system to the particular domestic conditions. He cautioned that expatriates working with extension programs in other countries "should be sensitive to cultural traits that bias counterparts, extension clientele, and curriculum content" (Diamond 1994, p. 76).

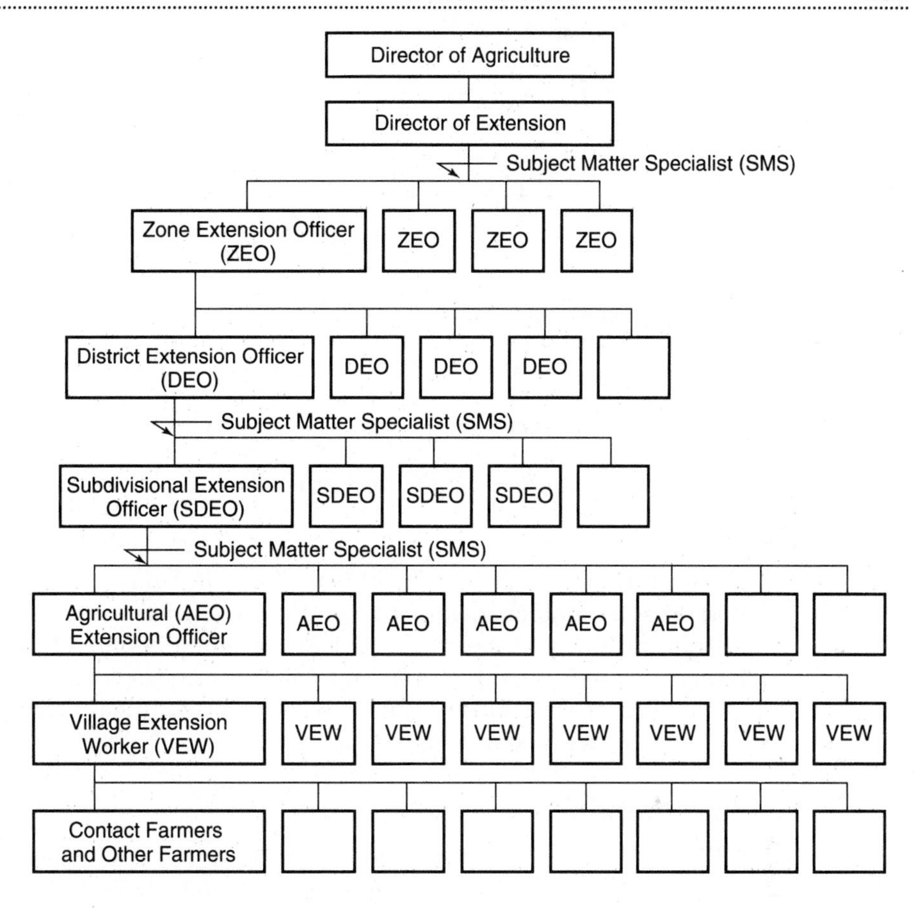

Figure 10-5 Organizational Structure of the Training & Visit System.

Project Approach

The Project Approach uses technology transfer within the broader concept of integrated rural development. The central government controls program planning. Foreign funding and foreign advisers are characteristics of this approach. A concern is that when the funding ends, the project tends to stop. Also, there is typically "little if any coordination among donor programs in the actual countries" (Rivera, Seepersad, and Pletsch 1989).

Farming Systems Research and Extension

The Farming Systems Research and Extension (FSR/E) approach uses on-farm research. Scientific researchers use actual conditions rather than experiments in which all the variables are controlled. FSR/E requires strong commodity research support and researchers with good communication skills and travel funds to visit with farmers. Ideally, in this model, farmers

T & V (TRAINING AND VISIT)						
	M	T	W	Th	F	S
Week 1	T	V	V	V	V	E
Week 2	M	V	V	V	V	E

T = Training
Presentation by a team of Subject Matter Specialists of specific recommended practices to be demonstrated and discussed during the next two weeks.

V = Visit
Each visit within a two-week period is to a different farmer or group of farmers. Often these are contact farmers who in turn are expected to reach others. The contact farmers may be volunteers.

E = Extra Visit

M = Meeting
A supervisory meeting by the Area Extension Officer with a group of 20 or 30 Village Extension Workers.

Figure 10-6 An Example of T & V Scheduling.

test, evaluate, and diffuse information informally. However, the FSR/E approach "enables social and technical scientists to work together to learn from but not with farmers" (Lev and Acker 1994, p. 42). In the Participatory Action Research (PAR) approach, there is a "three-way collaborative learning process among social scientists, technical scientists, and farmers" (Lev and Acker 1994, p. 42).

Participatory

In the participatory model of extension as well as the participatory model of research, activities are conducted though organizations such as farmers' associations and other groups. Examples are East African farmers' clubs organized by extension that serve as the main delivery system for credit to farmers.

> *Clubs decided how the credit would be distributed and also assumed collective responsibility to repay farmers' loans. An almost unparalleled record of close to 100 percent repayment was achieved, at least in part the result of the policy that denied a defaulting club further credit (Casley and Kumar 1987, p. 144).*

Extension in India

India has had a highly organized extension service for almost half a century. It was one of the first countries to successfully implement the Training and Visit System and is known for its innovations in rural development programs. Agriculture in India has made tremendous strides since 1966 and the "Green Revolution," the increase in grain production due to better seed varieties. It achieved self-sufficiency in food grains in the mid-1980s with 80 percent of Indian households getting at least two meals a day. Extension efforts in food production and food storage strategies have helped keep India free of famine (Narasimha et al. 1992).

One of India's concerns is to reach the rural poor, and there has been an increased focus on reaching women who are involved in dairy operations (Smith-Sreen and Smith-Sreen 1992). There is a need for a holistic, systems approach if projects such as dairying are to be successful. Figure 10–7 illustrates the interrelationships of the whole rural system in successful dairy development. For example, proper training will not ensure success unless there is sufficient labor to take care of the goats and a market for the milk.

THE CHINESE MODEL

The Chinese model is similar to the developing country approach in that it includes both education and regulation , but there are some differences in its methods and approaches. It is a model characterized by strong central direction and divisions by commodity, such as rice or beans.

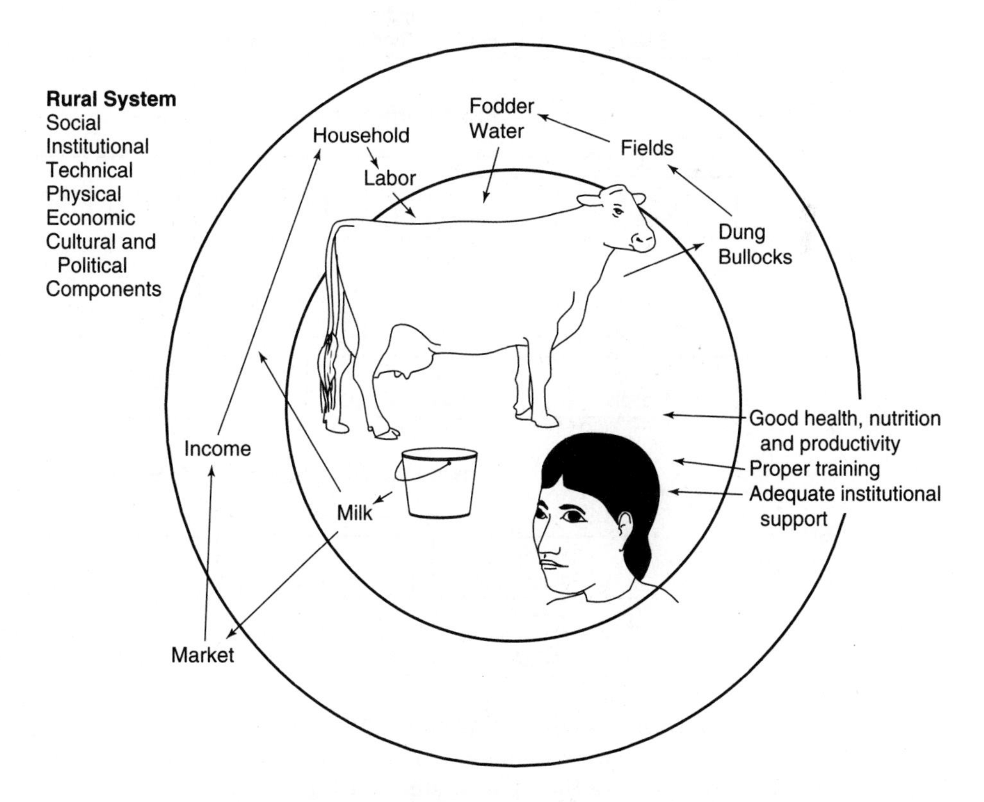

Figure 10–7 Strategies for Successful Dairy Development. Poonam Smith-Sreen and John Smith-Sreen, Department of State Libreville, Washington, D.C.

Extension in Mainland China

Mainland China (People's Republic of China) is a vast country with almost two-thirds of the world's extension employees (Bartholomew 1994). There are extension organizations at all levels of government, each fairly autonomous, with annual agreements and funding arrangements signed between levels. Figure 10–8 illustrates the complexity of the government bureaucracy in the field crop division. Ou, Xiaoyun, Yonggong, Dehai, and Xiaoying (1995) suggest a need for a multi-disciplinary extension service rather than the traditional one based on single disciplines. They advocate

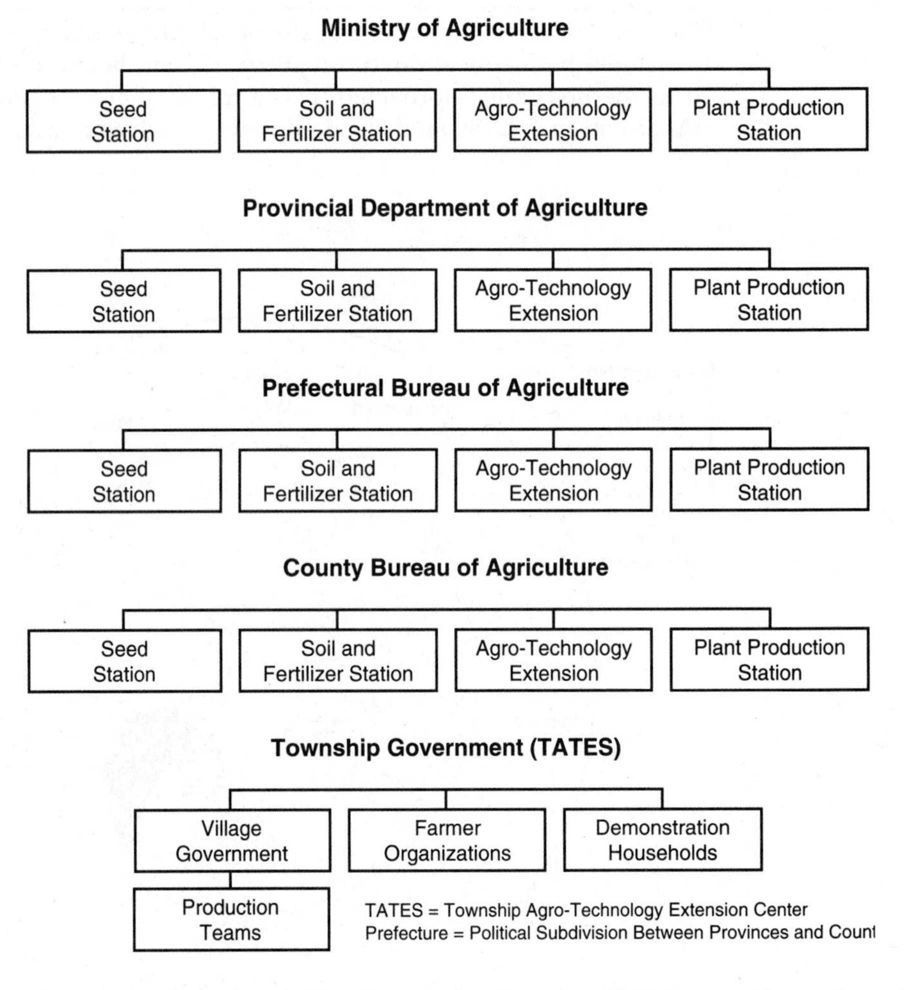

Figure 10–8 *Chinese Agricultural Extension Agencies: Field Crops. Chart adapted from Delman, J. 1991.* **Agriculture Extension in Renshou County, China: A case study of bureaucratic intervention for agriculture innovation and change.** *Unpublished doctoral dissertation, University of Aarhus, Denmark.*

a stronger role for agricultural universities in agricultural extension as China moves toward a market economy. There is a real need for professionally trained extension staff.

Part-time farmer technicians are employed at the village level, and extension delivery methods include village loudspeakers, blackboards, songs, dance, and slogans (Foo 1992). Signed contracts between the farmer and extension for yield goals result in reimbursement to the farmer if yield falls below a certain percentage. If yield exceeds the goals, extension receives a bonus, providing a strong incentive for staff to provide the best supplies and services to their clients. (Food and Agriculture Organization 1991).

Extension education in China is a complex, multi-ministry structure and includes research institutes, organizations for women and youth, and demonstration households. The research institutes operate at several levels and typically disseminate their research findings. The All-China Women's Federation offers educational programs, and the Communist Youth League for those fourteen to twenty-eight years of age provides educational activities for members and nonmembers. Demonstration households are a popular extension method (Delman 1991).

Following market reforms, the extension system in China was greatly decentralized and was expected to serve 170 million family farms rather than 6.5 million production teams. This was a great challenge for extension staff who were used to following centralized orders for technology delivery and adoption (Dehai and Yonggong 1994).

Extension in Taiwan

Taiwan has a successful system that effectively links research and farmers. Rather than having research, Extension, and resident teaching under one management such as in the U.S. land-grant universities, it assigns these roles to different agencies with provision for coordination. The extension role is further divided at the local level between public extension advisers and farmers' association advisers. There are interactive arrangements to ensure personal, group, and media contacts. These provide farmers with a continuing supply of current farm information.

EXTENSION IN WESTERN EUROPE AND CANADA

In the developed European countries, extension is considered to be an advisory service, and producers are often expected to pay for the services. Within the ministry of agriculture, each crop has its own section, and extension personnel are knowledgeable only about their own particular section. The advantage is that personnel can focus on just one subject, such as wheat. The disadvantage is that people may need information or assistance across a broader range of topics. Seed and fertilizer dealers, banks, producer cooperatives, and processing and marketing firms all provide advisory services to the clientele. The universities may train extension workers, conduct research, and develop materials, but they are not involved in advising extension clientele (Hoffman 1992).

The Netherlands

An example of European extension efforts may be seen in the Netherlands. Dutch agriculture may be the most productive in the world: in 1988 one farmer there fed 112 people compared to 85 in the United Kingdom, 79 in the United States, and 44 in France (Roling 1988). Dutch farmers are organized into groups, participate in policy formation, and pay half the costs of adaptive research. Computer networks of small groups of farmers exchange information and study each other's management practices.

Canada

Extension in Canada is similar to that in the United States in that it has a distinguished record of service in agriculture, home economics, and youth work. The main difference is that Canadian universities usually do not have any responsibility for extension. Extension, along with public education, is a provincial responsibility, and the provinces vary in their structures and programming emphases (Blackburn 1995).

Canada distinguishes between the public, private, and voluntary sectors of extension. Public-sector extension is paid for by government funds; private-sector extension is paid for by fertilizer companies, financial agencies, or other profit-oriented firms. Voluntary-sector extension is that paid for by farm organizations, commodity groups, or other organizations and individuals (Blackburn 1995).

There is a trend toward privatization in France, the United Kingdom, and the Netherlands, as well as in Canada. Extension may charge for services to individuals that directly result in improved incomes. Combined public-private funding is appropriate for applied research, training of farmers and agents, and improvement in extension methods.

EASTERN BLOC AND FORMER USSR COUNTRIES

The countries of the Eastern bloc and the former USSR are undergoing rapid changes in their systems of producing food and fiber as they dismantle the huge collective farms. In many cases, they are searching for an extension model on which to build an informational system.

In the Eastern European region, profound changes in policies are taking place with increased privatization of goods and services (Hoffman 1992). These policy changes will have a major impact on the development of extension structures and missions in the countries of the former USSR.

For example, in the Czech Republic, land is being returned to people who owned it in the early 1900s. These nonfarmers need help with basic skills in production as well as management of an agricultural enterprise. Lack of resources to do such basic tasks as remove parasites from sheep hinders production. Emphasis is on increasing production to feed the populace with little interest in extension's role in family, youth, and community development.

PROBLEMS AND CONSTRAINTS

Some of the problems faced by extension are the same the world over. Others are problems faced mainly by developing countries in Africa and parts of Asia and Latin America.

Constraints in Developing Countries

There are many barriers to effective extension education in developing countries.

1. Poorly trained, poorly paid personnel who lack transportation. Even if transportation such as motorcycles is provided, workers may not use them for fear of attacks by dogs (Roling 1992).
2. Regulatory and input-dispensing duties that impinge on time spent on educational programming. Inputs include seeds, fertilizer, and credit that are supplied by extension rather than the private sector. The availability of inputs from the private sector tends to be low when the policy and regulatory environment is poor, when populations are remote from urban areas, when passable roads are lacking, and when production consists mainly of basic food by subsistence farmers (Raman 1992).
3. Advice that is poorly timed.
4. Illiterate clientele. Clients who do not know how to read may be difficult to reach with new ideas. In a study of Turkish villagers, Lerner (1981) found that literates were more responsive to innovation and more likely to take initiative in modernization activities than were illiterates in the same village.
5. Counter-productive governmental policies, wars, drought, civil unrest, and need for land reform.
6. Social and cultural factors that make female audiences difficult to reach. For example, of the total number of trained extension personnel in twenty-one African countries in the early 1990s only 3.4 percent were women (Asiabaka 1992).
7. Lack of monitoring and evaluation. Monitoring involves inspecting periodically to see if people are doing what they are expected to do, while evaluation is a broader concept that determines the worth of a program. Monitoring typically takes place at the lower levels of Roling's model, see figure 10–9, but a program's worth becomes evident at the higher levels. It is easier to monitor (compare inputs and outputs) than it is to evaluate (assess effects and impacts). Two effects in need of evaluation are: (1) articulation

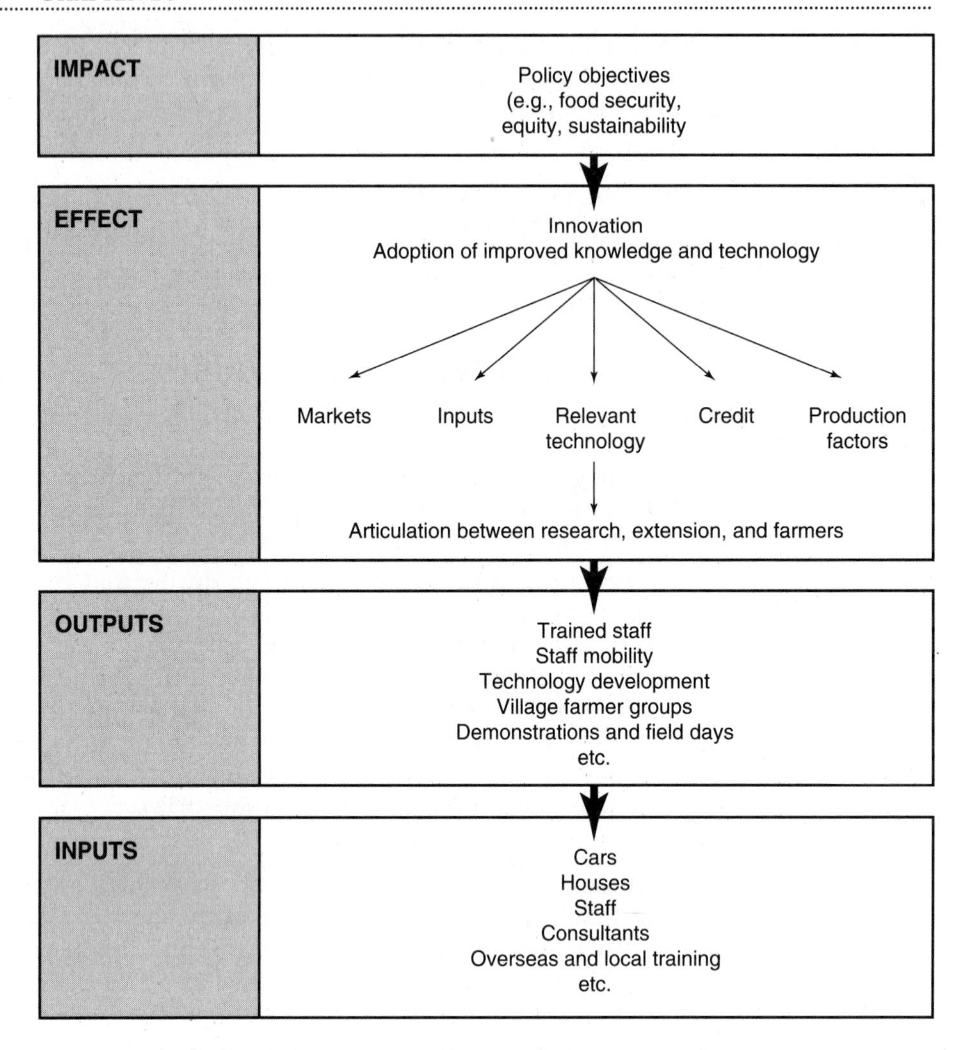

Figure 10–9 Framework for Monitoring. Niels Roling, Wageningen Agricultural University, the Netherlands.

between research, Extension, and farmers and (2) the access of farmers to markets, credit, and technology (Roling 1992).

8. Poor selection of local leadership. A study of demonstration farm operators in Finland found that the extension-selected farmers were not always the ones whom their neighbors asked for advice. Some of them did not want their neighbors to know that they were demonstration farmers, and people were more likely to ask someone other than their nearest neighbor (Westermark 1981).
9. Advice that is not research based. Sometimes this is because the research is not available. At other times, it is because there are no in-

centives or structure to link research and extension efforts. Interactions between educational institutions and rural communities should benefit both the institution and the rural communities. But it is difficult to link research at a university with rural communities. A study by Quispe, Gamon, and Miller (1994) of a graduate college in Mexico found that professors at the main campus participated infrequently in the regional centers located in the rural areas. College and university structures tend to value research and teaching over extension activities (Hoffman 1992). European extension directors rated their research linkages as strong or very strong but African directors rated theirs weak or very weak (Swanson et al. 1990). This may be because most European extension services were started before World War II and had well-established agricultural research at universities. African extension services tended to be established after World War II and agricultural universities were weak or nonexistent.

10. Environmental concerns. In a study of extension in Peru (Brown 1992), natural resource conservation measures were ranked the number one need for content dissemination. This was surprising in light of the many critical problems facing agriculture in Peru.

Suggestions for Meeting Constraints

Studies are needed to determine which models best fit a particular situation in a particular country. An example of the kind of study needed was done by Hassanullah in Bangladesh (1992). He compared the performance and costs of the Training and Visit, the Advisory Service, and the Integrated Model (a new Sugarcane Extension Service model with combined features). Under the Integrated Model, staff were better paid and more experienced.

A possible new model for technology development was suggested by Acker, Marcey, and Bunderson (1992). The model in figure 10–10 makes use of knowledge that the indigenous people have developed to help them cope with local weather patterns, soil types, and cultural situations. The model's emphasis on problem-solving teams can help solve two constraints: poorly-timed advice and advice that is not research based.

Constraints in Industrialized Countries

Industrialized countries have some of the same problems as developing countries, such as rural-urban migration and environmental concerns. However, the main concern for countries that have had an exclusively agricultural extension system is the decreasing number of people involved in agriculture and the future of a system devoted in the past mainly to technology transfer of agricultural innovations. If we are to see extension prosper, it needs to expand its role as a nonformal educational institution that empowers people to make their own decisions. For exam-

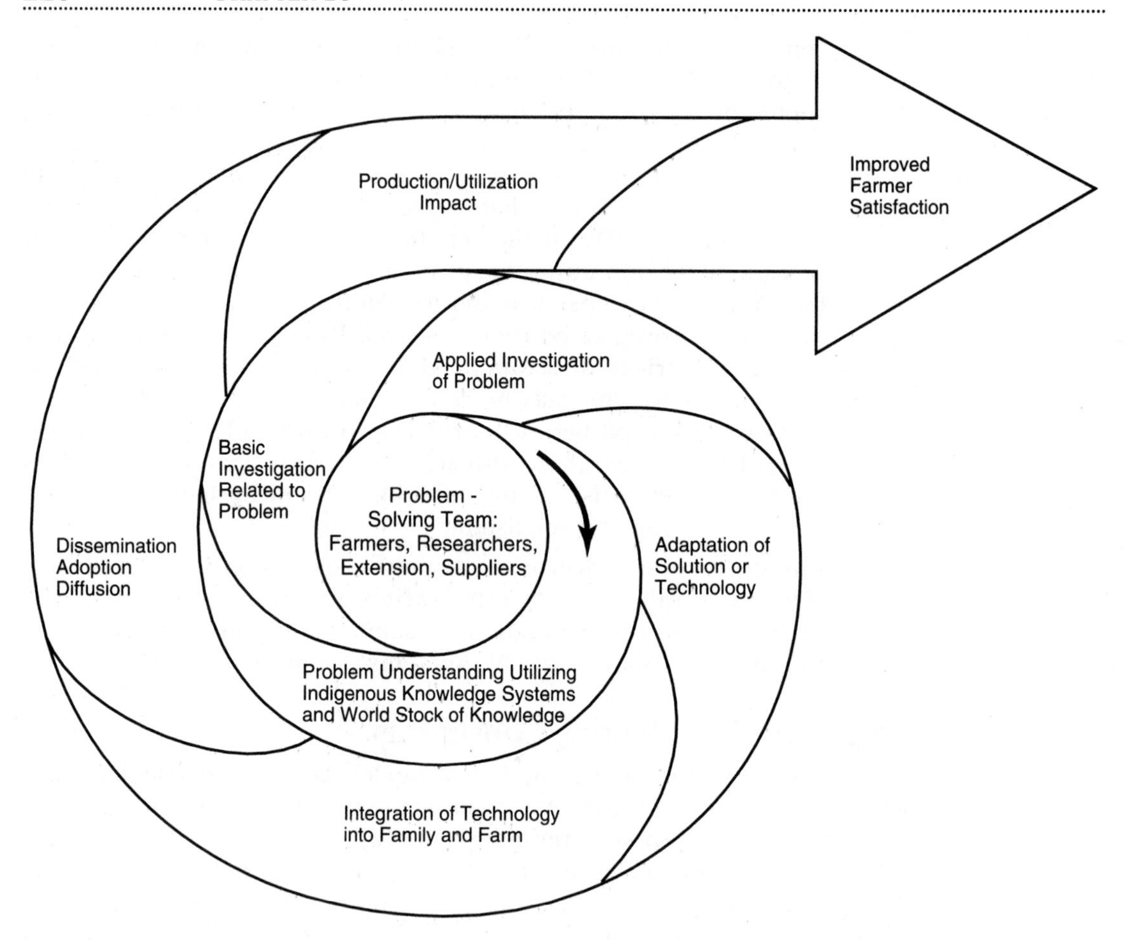

Figure 10–10 ***Proposed Technology Development Model.*** *Acker, D. G.; Marcey, J. L.; and Bunderson, W. T. 1992. New paradigms for technology transfer. In* **Proceedings, Symposium for research in agriculture and Extension education.** *Columbus, OH: Ohio State University.*

ple, in his study of Taiwanese vegetable farmers, Su (1993) cited a crucial need for extension to pay attention to leadership development.

Bennett (1992) recommended that extension emphasize education more than technology transfer. Rather than simply transferring and interpreting information on new technology, extension needs to provide "sequential short courses covering bodies of scientifically-based principles and processes."

For both domestic and international extension, intermediate users are an increasingly important audience. These intermediate users include both public-sector and private-sector entities. The public sector includes other governmental agencies; the private sector includes all those who provide inputs, such as banks and dealers in feed, seed, and fertilizer (Bennett 1992).

SUMMARY

Three reasons why it is important for people in the United States to study about extension in other countries are international opportunities, gaining ideas from other systems, and helping clients develop an international perspective. The opportunities include short-term experiences abroad, hosting international visitors, and careers. There are lessons to be learned from other extension systems; some countries have had successful programs for a long time. Three reasons that clients need an international perspective are the interdependence of our economies, environmental concerns, and the preservation of the gene pool for agricultural products.

Donor agencies are an important part of international extension, and the United States provides a major part of their funds. Donors include nongovernmental agencies (NGOs) and USAID as well as The World Bank and other United Nations groups.

Extension structures differ from country to country, but almost none follow the U.S. land-grant system of university responsibility and local funding. The Training and Visit System is operating in close to half of the world's developing countries and consists of a structured schedule of both training and visits to farmers. The Mainland Chinese system is the world's largest and is a multi-ministry, vertically organized system dependent upon negotiations between levels of bureaucracy.

All extension organizations are involved with information transfer and serve as change agents, and some challenges, such as environmental concerns and budget constraints, are pervasive worldwide. Extension systems in other countries differ from U.S. extension mainly in their structural organization and top-down philosophy. Also, in many countries programs focus on agriculture to the exclusion of family, youth, and community extension programs.

DISCUSSION QUESTIONS

1. Of the three reasons to study about international extension, which is more important now? In the future? What are other reasons to study international extension?
2. When donors make decisions on funding, what aspects of international programs might they consider?
3. Which features of the U.S. land-grant system might be more appropriate in a European country than in a developing country?
4. What kind of assistance should the United States provide to extension in developing countries: Surplus food and commodities? Teaching

equipment and supplies? Vehicles? Fertilizer and pesticides? Research, teaching, and extension personnel? Scholarships for students, researchers, teachers, and extension workers from other countries to come to the U.S. to study? Why or why not?

5. What modifications would the Training and Visit System need before it could be used in the United States?
6. What are the reasons for the preeminence of top-down programming in developing countries? What strategies could be used to increase bottom-up programming? What is there in the U.S. land-grant extension approach that has fostered local involvement?
7. Which plays a more important role in the continual process of increasing agricultural productivity: Research? Teaching? Extension?

REFERENCES

Acker, D. G.; Marcey, J. L., and Bunderson, W. T. 1992. New paradigms for technology transfer. In *Proceedings, Symposium for research in agriculture and Extension education.* Columbus, OH: The Ohio State University.

Amanor, K., and Farrington, J. 1991. NGOs and agricultural technology development. In *Agricultural Extension: Worldwide institutional evolution & forces for change*, W. M. Rivera and D. J. Gustafson, eds. New York, NY: Elsevier Science Publishing.

Asiabaka, C. C. 1992. Assessment of the training needs and job performance of women agricultural Extension personnel in Nigeria. *Journal of Extension Systems* 8(1):158–66.

Axinn, G. H. 1988. *Guide to alternative extension approaches.* Rome: United Nations Food and Agriculture Organization.

Bartholomew, H. M. 1994. The world's largest extension system. *Journal of International Agricultural and Extension Education* 1(1):32–36.

Bennett, C. F. 1992. A new interdependence model: implications for Extension education. In *Proceedings, Symposium for Research in Agriculture and Extension Education.* Columbus, OH: The Ohio State University.

Blackburn, D. J., ed. 1995. *Extension handbook.* Toronto, ONT: Keith Thomas Publishing.

Brown, R. 1992. An assessment and analysis of agricultural Extension programs in Peru: implications for agricultural development. In *Proceedings, Symposium for Research in Agriculture and Extension Education.* Columbus, OH: The Ohio State University.

Burger, A. 1994. *The agriculture of the world.* Brookfield, UT: Ashgate Publishing.

Casley, D.J., and Kumar, K. 1987. *Project monitoring and evaluation in agriculture.* Baltimore, MD: The John Hopkins University Press.

Cooperative States Research Service. July 1993. *Dynamics of the research investment: issues and trends in the agricultural research system.* Washington, D.C.: Cooperative States Research Service, United States Department of Agriculture.

Dehai, W., and Yonggong, L. 1994. Linkage analysis in agricultural extension: in the context of a sustainable rural development—the case of PR China. *Journal of Extension Systems* 10(2):47–59.

Delman, J. 1991. *Agriculture Extension in Renshou County, China: A case study of bureaucratic intervention for agriculture innovation and change.* Unpublished doctoral dissertation, Aarhus, Denmark: University of Aarhus.

Des Moines Register. May 25, 1994. *Look south, Iowans told.* Des Moines, IA.

Diamond, J. E. 1994. Agricultural extension in Swaziland: An evolution. *Journal of International Agricultural and Extension Education* 1(1):71–77.

Food and Agriculture Organization. 1979. *Agriculture, Towards 2000.* C79/24. Rome: United Nations FAO.

_______.1991. *Preparation report for agricultural support services project.* A report prepared for the World Bank. Rome: United Nations FAO.

Foo, F. May 1992. *Strengthening agricultural extension in China and FAO's involvement.* Paper presented at International Seminar on Important Trends in Agriculture Extension, Beijing, Peoples' Republic of China.

Greenland, D. J. 1990. Agricultural research and third world poverty. In *Developing World Agriculture*, Speedy, A., ed. London: Grosvenor Press.

Hassanullah, M. 1992. Organizational properties and manifestations of different models of extension work. *Journal of Extension Systems* 8(1):80–102.

Hayward, J. A. 1990. Agricultural extension: The World Bank's experience and approaches. In United Nations Food and Agriculture Organization (FAO) Report: *Global consultation on agricultural extension.* Champaign, IL: University of Illinois, Champaign-Urbana. INTERPAKS.

Hoffman, H. K. F. May 1992. What does the future hold for higher agricultural education? In *Proceedings, Symposium for research in agricultural and Extension education.* Columbus, OH: The Ohio State University.

Lerner, D. 1981. Literacy and initiative in Turkish village development. In *Extension education and rural development* Vol. 1, B.R. Crouch and S. Chamala, eds. New York, NY: John Wiley & Sons.

Lev, L. S., and Acker, D. G. 1994. Alternative approaches to technology development and adoption in agriculture. *Journal of International Agricultural and Extension Education* 1(1):37–44.

Mohamed, I. E., Gamon, J. A., and Trede, L. D. 1994. Philosophy, mission, and focus of agricultural extension in Africa, Asia, and Latin America. *Journal of Extension Systems* 10(2):26–35.

Mohamed, I. E., Gamon, J. A., and Trede, L. D. Fall 1995. Third world agricultural extension organizations' obligations toward the educational needs of rural people: A national survey. *Journal of International Agricultural and Extension Education* 2(2):11–17.

Narasimha, N., Ranganath, M., Miller, L., and Chandrakanth, M. 1992. New instruments in extension education for imminent environmental problems of Indian agriculture. In *Proceedings, Symposium for research in agricultural and Extension education*, Columbus, OH: The Ohio State University.

Ou, L.; Xiaoyun, L., Yonggong, L., Dehai, W., and Xiaoying, J. 1995. The emerging roles of the universities and colleges in agricultural extension and rural development in China. *Journal of International Agricultural and Extension Education* 2(2):68–78.

Quispe, A., Gamon, J., and Miller, W. 1994. Impact of the regional centers on faculty and students at the Colegio de Postgraduados, Mexico. *Journal of International Agricultural and Extension Education* 1(1):63–70.

Rajasekaran, B., and Martin, R. A. 1990. Evaluating the role of farmers in training and visit Extension system in India. Paper presented at the *Tenth Farming Systems Research/Conference.* East Lansing, MI, October 1990.

Raman, K. V. 1992. Agricultural and extension education: Preparing for the 21st century. In *Proceedings, Symposium for Research in Agricultural and Extension Education.* Columbus, OH: The Ohio State University.

Rasmussen, W. D. 1989. *Taking the university to the people: seventy-five years of Cooperative Extension.* Ames, IA: Iowa State University Press.

River, W. M.; Seepersad, J., and Pletsch, D. H. 1989. Comparative agricultural extension systems. In *Foundations and changing practices in Extension*, D. J. Blackburn, ed. Guelph, ONT.: University of Guelph.

Roling, N. 1988. *Extension science: Information systems in agricultural development.* Cambridgeshire, NY: Cambridge University Press

Roling, N. 1992. Effects of applied agricultural research and extension: issues for knowledge management. *Journal of Extension Systems* 8(1):166–183.

Smith-Sreen, P., and Smith-Sreen, J. 1992. Research and extension needs of Indian women dairy farmers: implications for agricultural and extension education. In *Proceedings, Symposium for research in Agricultural and Extension Education,* Columbus, OH: The Ohio State University.

Su, Y. F. 1993. Utilizing evaluation to develop Extension strategy for vegetable farmers in Taiwan. In *Proceedings, Extension education evaluation topical interest group, American evaluation association.* Washington, D.C.: Planning, Development, and Evaluation Staff, Extension Service, United States Department of Agriculture.

Swanson, B. E., Farner, B. J., and Bahal, R. 1990. The current status of agricultural extension worldwide. P.43-75 in United Nations Food and Agriculture Organization (FAO) Report: *Global consultation on agricultural extension.* University of Illinois, Champaign, Urbana, IL: INTERPAKS.

Westermark, H. 1981. Demonstration farmers' role in the adoption of innovations. *In Extension education and rural development.* vol. 1, B.R. Crouch and S. Chamala, eds. New York, NY: John Wiley & Sons.

Wisner, R. N., and Wang, W. 1990. *World food trade and U.S. agriculture, 1960-1989.* 10th ed. Ames, IA: Midwest Agribusiness Trade Research and Information Center.

World Bank, The. 1990. *Agricultural extension: The next step.* Washington, D.C.: Agriculture and Rural Development Department.

World Factbook, The. 1995. Washington, D.C.: Central Intelligence Agency.

CHAPTER

> *"My interest is in the future, because I am going to spend the rest of my life there."*
> —CHARLES KETTERING

Extension Future Focus

KEY CONCEPTS

- Societal trends
- Assessments of Extension efforts
- Future focus
- The challenge ahead

EXTENSION IN A COMPLEX WORLD

We have become so used to free public schooling for children that we forget it was a radical idea in the 1800s. Now it is accepted without question that we need to educate every child. Some have also asked if we need a national nonformal educational system. We already have such a system; its called the Cooperative Extension System. As our world becomes more complex, the need increases for a trusted, capable, nonformal public educational system to complement the formal school system.

> *The wealth and competitiveness of a nation depends increasingly on the extent to which its citizens are educated. Economies are becoming less dependent on natural resources and more dependent on people's intellectual and interpersonal skills. Education equips people to continuously identify what they need to know and to learn it quickly. Today's economic transformation toward a global economy requires further public investments to educate users including enabling them to define problems, question assumptions, test alternates, and organize to solve problems.*
>
> (BENNETT 1994, P. 41).

Today the need for Extension to deliver production agriculture information directly to farmers is decreasing, but there are other societal issues in need of educational efforts and there will always be problems that education can help solve.

A complex world has many different, interrelated parts that are continually changing in importance and function. Jobs as we know them may cease to exist; instead there may be tasks or functions. The new vision is of a world filled with different groups of people and new family structures, information systems, and production processes. Early childhood education is not likely to serve for a lifetime, so work skills will need continual updating. Skills in human interaction will loom large in importance. The future is one of contradictions in values and priorities; many will be uneasy about their roles and the expectations of others. Education will be a vitally important role in helping people to update their work skills and adjust to societal issues to come.

The Extension Service has helped people adjust to the many changes that have impacted American society in the twentieth century, such as the depression, World War II, the technological revolution, the emergence of biotechnology, and changing family economics and lifestyles. A complex world is interesting and challenging and one to which Extension can bring a welcome message of hope. Extension programs will continue to offer ideas for better living, enhanced skills for personal empowerment, and practical education that promises economic and social betterment. Extension programs are most effective when they are based on societal needs and accurate assessment of those needs. Although no one can predict the future, this chapter will address some societal trends and Extensions' opportunity to respond to those trends and issues. Also discussed will be issues, challenges, and concerns within Extension that will shape future directions and programs.

SOCIETAL TRENDS

What are the societal trends that will become societal issues? What will future communities look like? What educational delivery methods will be appropriate? What will be the content of the programs? Answers to these questions are based on the work of futurists, people who have made careers of observing current trends and predicting the future, in both sociological and economic terms.

Eric Toffler made "futuring" a household word with his book *Future Shock* in 1970. Twenty years later in *Power Shift* (1990), one of several trends he mentioned was the aging of America and an increase in older workers. Other trends Toffler noted were an increase in the minority population, the continual elimination of job titles, and the introduction of new jobs necessitating retooling and retraining. Naisbitt and Aburdene (1990) also identified new directions for the 1990s in the best seller *Megatrends 2000*, including global economics, cultural nationalism, privatization of the welfare state, more women in leadership roles, the "age of biology," and the triumph of the individual.

The approach of a new millennium has caused many to speculate about what the future holds. The United Way of America's Strategic Institute (1989) identified more than a hundred specific trends in American society and grouped them into nine forces reshaping how Americans live and think. These nine forces—defined in terms of their social, economic, political, and technological impacts—summarize the works of almost all forecasts. The nine forces reshaping America described by the United Way Strategic Institute are:

1. **The Maturation of America:** aging "baby boomers," the "greying" of America
2. **The Mosaic Society:** diversity trends in education, ethnicity, age, and family structure
3. **Redefinition of Individual and Societal Roles:** public versus private sector, individuals taking on more responsibilities and relying less on large institutions (e.g., for personal health)
4. **The Information-based Economy:** changes in technology and information transfer
5. **Globalization:** movement of products, capital, technology, and information around the world
6. **Personal and Environmental Health:** quality-of-life issues
7. **Economic Restructuring:** global economic competition, deregulation, changing consumer preferences
8. **Family and Home Redefined:** changing family structures and functions (e.g., child care and food service being handled outside the home)
9. **Rebirth of Social Activism:** public agenda focused on social concerns

One indication of societal change is family structure. It is now increasingly rare to find a household that contains a traditional nuclear family, that is two parents and their biological children. "Empty-nesters," young single adults, single parents, and families with stepchildren are part of the 90 percent of households that do not consist of nuclear families (U.S. Census Bureau 1990). This change is important because changes in living arrangements precipitate changes in many other areas, such as spending patterns and attitudes toward education. Still other societal trends are the individualizing, or customizing of goods and services and the "vigilante consumer," a term coined by futurist Faith Popcorn (1991). The environmental movement and technological factors were among the social, political, and economic factors listed by the National Agricultural Extension Leadership Workshop in 1994. For some futurists, the list of basic necessities—food, clothing, and shelter—has expanded to include health care and education (Segal 1991).

Extension Responses to Societal Trends

In addition to its traditional or base programming, Extension has adopted specific approaches in response to these trends and issues (Boyle 1990). Three of the primary approaches—issues programming, national initiatives, and strategic planning were addressed in detail in Chapter 5.

The Government Performances and Results Act of 1993 has also had an impact on how the Extension Service has responded to issues and trends. This legislation requires that all agencies submit a five-year strategic plan that includes the following:

- A mission statement (major functions and operations)
- Outcome-related goals and objectives
- Strategies for achievement of goals and objectives
- External factors affecting achievement
- Evaluation plan featuring numerical measures of impact

The Performance and Results Act required the development of strategic plans by 1998 and reports containing numerical measurement of the achievement of performance objectives compared to goals by the year 2000 (Office of Management and Budget 1993). This means that all government programs, including Extension, need to focus much more specifically on what they plan to do and how they will measure in numbers what they have accomplished. It means that baseline data, as well as measurements at the conclusions of programs, need to be collected in order to measure gains.

The first step is the development of the plan. By definition, strategic plans include an assessment of external forces (Pfeiffer et al. 1989). We call forces that are outside the organization "environmental forces" and the process of identifying them "environmental scanning." Multiple scanning methods, including visioning teams, focus groups, and surveys of participants and nonparticipants should be utilized. The external forces uncovered by a national environmental scanning process may not fit a particular state or a local situation, therefore analysis of state and local situations is essential for program success.

ASSESSMENTS OF EXTENSION

Assessments of Extension by both internal and external reviewers is not new. In order to grow and develop to meet the changing needs of society, any organization needs to continually assess its strengths and challenges.

Clientele Satisfaction

Of significant importance to Extension is the level of satisfaction of the clientele being served. Extension's clientele base consists of people and communities throughout the nation that have an interest in and a need for information and services that Extension provides through its base programs and selected initiatives.

Warner and Christensen (1984) assessed the public's opinions nationally about Extension. They found that users' evaluations were more favorable than those of nonusers and recommended increasing the numbers of users and ensuring a high level of satisfaction among them. A national survey in 1995 (Christensen et al.) found that after hearing a brief description, 85 percent of the respondents indicated that they had heard of the Cooperative Extension Service before being interviewed. Of that same group, 26 percent had used the programs or services of CES. Strong support for the Cooperative Extension Service was evident when more than half of the study participants indicated that more tax dollars should be used to support Extension programs.

Many states and local units have conducted their own assessments with results similar to the national study. In Oregon (Kingsley 1988), 98 percent of users and 89 percent of nonusers said Extension was a good source of information. In Texas (Herren 1995), a carefully randomized telephone survey that compared attitudes toward Extension of metropolitan and nonmetropolitan respondents predictably resulted in less favorable responses from city residents. This provides a strong indication that, especially in times of increased scrutiny under tight budgets, Extension needs to bring its messages to a larger, more diverse audience.

Internal Assessments

Since World War II, Extension has conducted major internal evaluations at regular intervals. The key points of four Extension evaluation reports summarized in the *Journal of Extension* by Ratchford (1984) are shown in figure 11-1. In 1991, another Extension summary was completed. The items set in boldface type indicate an addendum to Ratchford's summary of Extension reports to include the impact of the "Extension in the 80s" report and the 1991 summary.

Each decade report arose out of a need not only to assess the current situation but also to anticipate concerns and issues and to develop plans for addressing them. The decade reports addressed the scope and nature of the Extension Service, reviewing the basic mission and philosophy, programming content and approach, clientele base, funding sources, relationships, and staffing. The impacts have been briefly described after each report. Over the years, new program areas have been added (community development, natural resources), new programming approaches have been adopted (four-year plans, issues programming, national initiatives), the scope and mission have been constantly revisited and linkages have been reviewed, and there has been a strong demand for greater accountability and evaluation of program impact.

Each of these individual areas has also been the subject of research efforts related to Extension. A search of the literature about Extension topics reveals an extensive database describing characteristics, programs,

ITEM	STUDY				
	Joint Committee	**Scope Report**	**A People and A Spirit**	**Extension in the 80s**	**Patterns of change**
Year printed	1948	1958	1968	1983	1991
Motivations	Return to normal after 20 years of emergencies.	Farm surpluses low prices, related agencies' concerns, and Sputnik.	Social programs of New and Great Societies.	Chronic questioning of role, funding, and control.	Societal changes Shifting demographics
Approach used	Joint committee appointed by Secretary of Ag. and NASULGC	ECOP sub-committee.	Joint committee appointed by Secretary of Ag. and NASULGC	Joint committee appointed by Secretary of Ag. and NASULGC	Strategic Planning Council Futuring panels
Key issues	All addressed scope of subject matter, clientele, methodology, training, financing, and relationships. The basic philosophy was restated in all. There was consistent broadening of program scope. The unique features were:				
	Relationship with farm organizations and emphasis on training.	Broadened program scope by emphasizing management, marketing, and public policy.	Strong emphasis on social programs and the disadvantaged. Brought 1890 institutions into the picture.	The emphasis placed on the partnership of USDA, universities, and local government.	Issue-based programming Use of new technology.
Impact	Changed relationship with farm organizations, stronger tie of Extension to academic base, and new training opportunities.	Program areas other than ag./home econ. and 4-H became part of system.	Less than others because of little follow-up. Did increase visibility of social issues.	Four-year Plans Environmental programming	Reorganizing and restructuring Partnerships and collaboration

Figure 11–1 ***Summary of key points of four Extension evaluation reports. Amended from Ratchford, C.B. (Sept./Oct., 1984). Extension unchanging, but changing.*** **Journal of Extension.** ***31(3):17–19.***

attitudes, and perceptions leading to recommendations for change and improvement.

The most recent internal study assessing the Extension Service surveyed Extension administrators or directors nationally (Warner et al. 1996). National trends in Extension revealed by the survey can be summarized as follows:

1. **Positioning of Extension within the Land-Grant University**
    - Although there has been considerable discussion concerning the movement of Extension out of the college of agriculture to more universitywide outreach units, 74 percent of all Cooperative Extension Services still remain in agriculture.

- However, 28 percent of those surveyed indicated they had experienced changes in that arrangement in the past five years or are anticipating one in the near future.
- There is widespread support for the view that Extension programs need to be able to draw upon the total university's expertise.

2. **Securing Program Expertise**
   - 56 percent of the institutions believe that support of Extension is not viewed as a university expectation.
   - Expertise needed from academic units other than the traditional areas of agriculture and home economics is predominately secured through informal contacts among faculty and staff or purchased on a contractual basis
3. **Influence of the Partners**
   - There is general satisfaction with the balance that exists between county and state partners; however more involvement of the local partner in areas of salary, accountability, and hiring is desirable.
   - More state and less federal influence in programming decisions is desired.
4. **Managers of Field Programs**
   - 71 percent of the institutions utilize middle-level managers in area or district offices. Most managers have both supervisory and programming responsibilities.
   - 74 percent of the institutions utilize county directors or chairs with administrative responsibilities ranging from 3 to 100 percent, with the average being 25 percent.
5. **Field Program Staffing**
   - The single-county staffing pattern remains the most prevalent (42 percent).
   - 30 percent of the institutions reported that they do not have a dominant staffing pattern, indicating that no single pattern describes at least 75 percent of their field staff.
   - The majority of county office sites are co-located with county government offices, although 38 percent indicated they have their own separate offices.
6. **Specialists Positions**
   - The greatest barrier to program flexibility is perceived to be faculty tenure. Although various patterns exist, the predominant approach is to offer tenure to specialists with Ph.D.s in the academic department of their specializations.
7. **Clientele Involvement**
   - 67 percent of all institutions reported that they have a "statewide" advisory group that provides input into determining the nature of the *overall* Extension program.

- Forty-one institutions reported that all of their counties have overall advisory groups, another fifteen indicated that more than half of the counties have boards, councils or advisory committees.

8. **Budget Process**
   - State monies designated for Extension generally are directly appropriated to Extension or handled as part of the university budget.
   - The perception exists that state support for Extension programs has not been appreciably influenced by how the institution is organized. However, there may be greater potential for future increases in a more broad-based university outreach effort.
9. **Capacity to Support the Six National Initiatives**
   - Greatest support was for the Children, Youth, and Families at Risk, and the Water Quality initiatives.
   - Medium capacity was identified for Sustainable Agriculture and Food Quality and Safety.
   - The lowest level of capacity was for Communities in Economic Transition and Decisions for Health.

In addition, when directors or administrators were also asked their perceptions about the issues or trends that are of most concern to the future of Extension, the following were identified.

1. **Funding:** state, federal, local, user fees, grants, anti-tax sentiments
2. **Clientele Needs:** changing demographics, economy, flexibility
3. **Organizational Influences:** university outreach, staffing and structural changes, downsizing
4. **Delivery Methods:** increased use of electronic delivery, multimedia technology, and user fees
5. **Support of CES and Higher Education:** lack of public support for higher education, declining support of traditional support groups, shift in political power from rural to urban, need to broaden the support base
6. **Mission:** shift from agriculture to urban focus, need for higher impact and more visible programs
7. **Partnerships/Competition:** need for more local involvement, collaboration with other agencies, change in federal government structure and positioning

FUTURE FOCUS

Identification of concerns and issues is not enough. Reflections and discussion must be translated into action for the process to have meaning. Identified below are a few of the issues the Extension Service faces and a discussion of some of the possible changes the organization might experience.

The Mission

The mission of the Cooperative Extension Service has been under scrutiny for many years. In 1989, when CES celebrated its seventy-fifth anniversary, many articles were published asking if a seventy-five-year-old mission could still be valid today. The biggest debate lies in whether the mission should be broad based to reach the masses or whether it should be narrowly focused and concentrate on doing a few things well. The significant consensus is that Extension's emphasis is on education; it should never lose site of the fact that its primary focus is to provide education to the people. Changes in how the mission are to be achieved is likely to focus on the empowerment of individuals to "solve their problems, realize their opportunities, and shape their own destinies" (National Workshop 1994, p. 17).

Program Planning and Evaluation

Changes in program planning and evaluation are expected to be in the direction of more strategic planning and more emphasis on issue programs. Staff will need to document changes in clientele practices using valid evaluation methods. They will also need to continue to utilize local advisory committees and support their recommendations and observations in conjunction with state and national initiatives.

The trend toward interdisciplinary efforts will continue, mirroring the interdepartmental trend at colleges and universities. Tompkins (1989) states

> *Extension can't serve American agriculture, let alone America, with relevant research-based knowledge and technology transfer using only the resources of Colleges of Agriculture as they're currently structured.*

There is much happening in the natural and biological sciences that significantly influences education about and in agriculture. Genetic research, biotechnology, food safety, and environmental issues are just a few examples. Other program areas are equally influenced and there is much that can be gained by utilizing resources of the entire university for programming related to the family, business management, communication, education, and public policy. The list is endless. In a time of limited resources, serious consideration needs to be given to utilizing all that is available. Extension can use its leadership and experience to initiate the broader outreach functions of the total university

Program Content

Changes in program content will likely continue to address the changing needs and issues of society. For example, increases in social activism may indicate more programming on public policy. Although Extension will continue to work with commodity groups and other special interest groups to help them become politically aware, Extension's role will continue to be one of facilitation and education and not promotion of any one political position.

Statistics tell us that only about 2.5 percent of the American population is in production agriculture. It is this group, however, that is still responsible for feeding and clothing the population. Issues in agriculture have changed from increasing production to sustainability in agriculture, genetic engineering, biotechnology, conservation of natural resources, food safety, and agricultural literacy. The wider variety of issues and concerns that Extension programs now address have value to audiences from all backgrounds and settings. Programming focus is on addressing needs and concerns not on place of residence or personal background.

The major program areas (agriculture, family and consumer, youth and community) that Extension has traditionally focused on will probably not change, but the issues and topics within those programs areas will evolve to more adequately address changes in the society at large.

Organization and Structure

Structural changes, such as staffing patterns and placement in the university, have been increasing since the 1980s. As seen in the 1996 report by Warner et al., changes are expected to continue. There will likely be more multicounty offices, fewer state specialists, more hiring of temporary personnel, and an increase in partnerships with community colleges. Clustering, in which staff from two or more counties work together, will continue, and county lines may be replaced by trade areas.

Still another change is the move toward interstate arrangements. State boundaries, which are artificial political divisions, may blur and give way to "micro-climate" partnerships, with specialists serving more than one state. An example of an interstate arrangement is the North Central Region Educational Materials Project that is a clearinghouse for bulletins and publications across the northern and western state region (Brown 1990). This approach to curriculum development lessens duplication and promotes sharing of resources.

Also anticipated is an increased use of intermediary clientele who will take information to the ultimate audience. Partnerships and new coalitions are going to be essential, not only to reach more people, but also to improve quality and results. Participants in Extension programs are increasingly educated, sophisticated, specialized clientele who have access to many different sources of information. As competition from private organizations offering the same services as Extension increases, there will be a growing opportunity for collaboration that enhances both public and private sources of knowledge.

Another change in structure is one we see occurring at the university level. There is a trend toward "university extension," a movement that goes beyond the traditional research base of agriculture and home economics. "Every faculty member should be regarded as a specialist whose skills may be brought to bear in developing and delivering programs.

Field agents must now serve more as information brokers, directing university expertise to areas of greatest need" (Yates 1994, p. 15). University extension uses the resources of the land-grant institution to meet the needs of the people, so the county office is truly a door to the university. The problem arises in the implementation. Some faculty are funded by Cooperative Extension; the ones who are not do not feel the same responsibility to deliver programs off campus. But it is imperative to "tap the knowledge base of the entire university" (Brown 1989, p. 5). Political adeptness will be needed to marshal the resources of various departments to answer the concerns of the land-grant university's supporters, the citizens of the state.

Funding

The 1995 study by Christensen et al. found that the public was not only aware of the Cooperative Extension Service but also felt that additional tax dollars should support this educational system. Many citizens, however, oppose any form of tax increase. The future is likely to hold less public funding for Extension and more funding from private sources. Fees for training programs, services provided, or materials are also likely to be enacted. Although some concern has been expressed about government agencies charging users' fees, precedent has already been set by tuition at state-supported universities and fees for visits to state and national parks and recreation areas.

Privatizing programs, using the model from the European community, may be a move for the future. In Britain, Holland, and Germany, clientele pay for advice on agricultural problems.

Another possibility is a change in the criteria on which funding is based and the processes for securing funds. Osborne and Gaebler (1992) outlined procedures for reinventing the government and suggested basing funding on outcomes, rather than on program inputs. Since the government is asking for evaluations that focus on outcomes rather than inputs, funding based on outcomes may soon follow.

Delivery Methods

Changes in technology have already altered delivery methods. The information highway makes the latest information immediately available to anyone with a modem and a computer. In an evaluation of an Extension grain marketing program, the information that state specialists shared through on-line data services was seen as more valuable than the content of newsletters. Farmers had already made their decisions by the time the newsletters arrived. Timeliness of information will be an important factor in future delivery method preference.

A nationwide study of farm profitability programs found that group workshops and meetings were the most common method of program delivery, followed by individual contact (Lippke et al. 1987). Meetings and one-on-one teaching, preferably on-site at the business, farm, or home, have long been favorites of Extension. Agents and specialists like the opportunity for immediate feedback and the variety that comes from a balance of office work and work out in the field. However, the days of face-to-face contacts as a primary delivery method have already diminished. Many more people can be reached by the new electronic methods than by the office calls and personal telephone contacts of the past. Distance education through fiber-optic networks, national satellite programs, computer networks, videos, compact disks, call-in radio programs, and toll-free telephone numbers will be strong supplements to the more traditional and existing methods.

Like most innovations, new delivery methods will offer some advantages and experience some limitations. Increased use of telecommunications may help hold down costs (Agnew and Foster 1991, Rockwell et al. 1993). Some methods have the advantage of anonymity. "A person who wouldn't attend a meeting or visit an Extension office might more easily and confidentially make contact and receive help from a telephone hotline" (Molgaard and Phillips 1991). Another advantage is that the new technologies can be used in interesting ways to enhance traditional deliveries. For example, Minnesota has widely distributed videos of its annual agricultural conference. Agents have augmented other programs with short video segments from the conference, and airings have occurred in meetings, schoolrooms, and on television (Stevens 1991). However, care must be taken to ensure that the new technology benefits all of society, rather than just those who have the resources to access the technology. There will always be a need for personal interaction. New delivery methods will serve as a supplement not a replacement to proven existing methods.

1890 and 1994 Institutions

The addition in 1994 of twenty-nine tribal community colleges to the seventeen traditionally black land-grant institutions (1890 institutions) in the southern states was an indication of further change by establishing new land-grant institutions that would predominately serve Native American populations. These were the first land-grant institutions to be established since 1890. As the percentage of minorities in the total population increases (Chesney 1992), the ability of the 1890 institutions to help reach that audience will increase in importance. The 1890 act provided for separate but equal institutions in those states that didn't permit blacks to enroll in the 1862 institution. The federal funds provided were supposed to be equitably, but not necessarily, equally divided between the two institutions in the seventeen states that had a second land-grant college. However, few, if any, institutions receive the same level of support as their 1862 counterparts. The 1890 and 1994 institutions have been more dependent upon federal funding than those founded in 1862. Cutbacks at the national level will prompt them to seek state and local funding (Hughes 1990) to survive.

Staffing

The backbone of the Extension Service is the staff employed to carry out its education functions and mission.

Some changes in staffing patterns have already been addressed. Although the single county pattern is still the most common, many states are using multiple staffing approaches. The projection is that more staff are likely to have multicounty positions (Agnew 1991). The "county" office of the future will probably serve at least two counties. Because multicounty staffing has been found to be the least satisfying arrangement for both agents and clientele, Bartholomew and Smith (1990), state that there is a need to experiment with a variety of patterns, such as clustering of counties, program assistants, and field specialists. Extension directors, however, indicate that the multicounty trend should be allowed to evolve locally rather than be imposed by the state (Warner et al. 1996, p. 5).

Other issues and trends will also influence staffing in the future. Budget constraints will force creative use of resources. Contracting or hiring individuals with special skills for short-term appointments may become necessary to ensure that issues are being addressed. Using such outside individuals with specific skills will allow the Extension professional to remain a generalist, but one with expert skills in accessing and utilizing resources. Use of electronic technology will be vital to access the many and varied resources available to assist clients with questions and problems. Ties to the university specialist and other resources will strengthen as assistance for problem solving becomes available quicker and easier. Economic constraints may also require stronger skills in networking and collaborating with other agencies and organizations as well as the development of a strong volunteer program that encourages delegation and empowerment.

Traditional skills of planning, organizing, program development, and evaluation and technical expertise may not be enough to fulfill the job responsibilities of the future. The future Extension professional will need to be an information expert, a forecaster and trend analyzer, be able to build a strong network and collaborate with others, and empower individuals and groups.

In *Megatrends 2000,* Naisbitt and Aburdene noted an increase of women in leadership roles. This will have an impact on Extension as more women slowly move in to administrative positions. Over half of the states and territories made their first appointments of women to county management positions in the 1970s, almost sixty years after the Extension Service began. But by 1989, there were still only eight (15 percent) women who held state director positions (Goering 1990).

THE CHALLENGE AHEAD

In 1995, the Extension Committee on Organization and Policy developed a new Strategic Framework to provide direction and support as Extension responds to opportunities, delivers programs, and makes decisions. In developing the strategic plan, the committee identified eight areas of tension or concern within the Cooperative Extension System. The following tensions, when viewed as challenges, will provide opportunities to shape the future direction of the Extension System:

1. **Role of Extension Professionals:** Extension professionals are both educators and information providers and must utilize both approaches as they establish learning partnerships with clientele.
2. **Rural versus Urban:** Extension must find ways to emphasize rural-metropolitan interdependence and serve audiences in both settings.
3. **Serving Diverse Audiences and Needs:** regardless of the location or background of its clientele, Extension must serve the needs of diverse audiences and focus on the critical issues to which it can contribute solutions.
4. **Extension and Research:** the research base is a unique characteristic of the Extension Service. Research agendas should be set collaboratively by the university and local communities.
5. **Relationships with Federal Agencies:** strong ties and collaborative relationships, not only with USDA but also with other related federal agencies, should be developed and maintained.
6. **Locus of Decision-making in Program Planning:** identification of programs based on need is a function of the Extension Service. Extension needs to value the contributions made by local advisory committees and constituents and carefully coordinate them with state and national initiatives.
7. **University Outreach:** the Extension Service has a unique role in the land-grant university and college. The Extension Service can use its experience and leadership as a model for the broader outreach functions of the entire university.
8. **Relationships among Land-Grant Institutions:** regardless of when an institution received land-grant status, opportunities for equitable funding and full participation must be available for all land-grant institutions.

SUMMARY

Extension is the largest nonformal educational agency in the world and is unique because of its lifelong thrust and its source of fresh, research-based knowledge. In a complex world, such a source of trusted, competent, public information is vital to an educated citizenry. Extension's mission today focuses strongly on empowering people to solve their own problems, rather than simply providing them with accurate, unbiased information.

Societal trends likely to influence Extension programs are the aging population, changing family structures, environmental awareness, and the desire for customized services. Extension has responded to these trends with issues programming, national initiatives, and strategic planning.

Extension's structure, program content, delivery methods, and personnel are likely to change in the future. The changes will be toward structures independent of political boundaries, content based on matters of wide public concern, electronic delivery methods, and personnel who are highly educated and experienced.

What will continue without change is the mission to help people improve their lives, a mission made possible by the dedication of the professionals, paraprofessionals, staff, and volunteers who are at the front-line of the vast educational effort called Extension.

DISCUSSION QUESTIONS

1. Do you agree that there is a critical need for a nonformal, lifelong public educational system? Why? Why not?
2. What aspects of the current structure of Extension would be useful in a framework for a general Extension system of service to all Americans? What aspects should change?
3. What are the main societal trends that will affect the future of Extension? What specific examples of these trends have you noticed?
4. What were the important features of Extension responses over the years to societal trends? In retrospect, which of these responses were most appropriate?
5. How is the content of Extension programming likely to change in the future? How would you describe the delivery methods likely to be used in the future? Pick a subject-matter topic and describe how it might be delivered ten years from now.
6. What changes in organizational structure are occurring in Extension now? How likely are these to continue? Why?
7. Are there changes other than those mentioned in the chapter that you foresee? Describe them and explain why they are likely.

REFERENCES

Agnew, D. M. Summer 1991. Extension program delivery trends. *Journal of Extension* 29(2):34–35.

Agnew, D. M., and Foster, R. Spring 1991. National trends in programming, preparations, and staffing of county level Cooperative Extension Service offices as identified by state Extension directors. *Journal of Agricultural Education* 32(1):47–53.

Bartholomew, H. M., and Smith, K. L. Winter 1990. Stresses of multicounty agent positions. *Journal of Extension* 28:10–12.

Bennett, C. F. June 1994. A new interdependence model: implications for Extension education. *Journal of Extension Systems* 10(1):33–41.

Bottom, J. S. July 1993. An assessment of state funding for the Cooperative Extension system. Washington, D.C.: Planning, Development and Evaluation Staff, Extension Service, USDA.

Bowen, C. F. June 1994. Job satisfaction and commitment of 4-H agents. *Journal of Extension* 32(1): Available: e-mail almanac@joe.ext.vt.edu send joe june 1994 research 2.

Boyle, P. 1990. The look of Extension in the future. *Poultry Science* 69:227–230.

Brown, N. Spring 1989. Too little, too late. *Journal of Extension* 27: 5.

Brown, S. Winter 1990. An easy search. *Journal of Extension* 28:37.

Chesney, C. E. Summer 1992. Work force 2000: Is Extension agriculture ready? *Journal of Extension* 30:30–31.

Christensen, J., Dillman, D., Warner, P., and Saleint, P. 1995. The public view of land-grant universities: results from a national survey. *Choices* (3):37–39.

Clark, R. W. Summer 1992. Stress and turnover among Extension directors. *Journal of Extension* 30:37.

Extension Committee on Organization and Policy. August 1990. *Conceptual framework for Cooperative Extension programming.* Washington, D.C.: Strategic Planning Council, Extension Committee on Organization and Policy and Extension Service, USDA.

_______.February 1995. *Framing the future: strategic framework for a system of partnerships.* A report prepared for ECOP and Cooperative State Research, Education, and Extension Service. Washington, D.C.: USDA.

Goering, L. A. Winter 1990. Women and Extension. *Journal of Extension 28:21–24.*

Harper, L. 1992. *The 21st Century Extension professional in the midst of organizational change:* Implications for motivational strategies. Unpublished master's thesis. Columbia, MO: University of Missouri.

Herren, J. 1995. Attitudes toward Extension of metropolitan and nonmetropolitan respondents. Proceedings, Southern Region Agricultural Education Research Meeting, Wilmington, NC.

Hughes, L. R. Winter 1990. The future of the 1890s. *Journal of Extension* 28:28–31.

Kingsley, K. K. 1988. *As others see us: a statewide survey.* Corvalis, OR: The Oregon State University Extension Service.

Lippke, L. A., Ladewig, H. W., and Taylor-Powell, E. 1987. *National assessment of extension efforts to increase farm profitability through integrated programs.* College Station, TX: Agricultural Extension Service.

Molgaard V. K., and Phillips, F. Winter 1991. Telephone hotline programming. *Journal of Extension* 29:13–16.

Naisbitt, J., and Aburdene, P. 1990. *Megatrends 2000.* New York, NY: Avon Books.

National Workshop. 1994. *Sustaining our future: A draft strategic plan for the agriculturally based program.* Washington, D.C.: National Extension Agriculture Program Leadership Workshop.

Office of Management and Budget. August 1993. *Implementation plan. Overview section. Government performance and results act of 1993.* Washington, D.C.: Office of Management and Budget.

Osborne, D., and Gaebler, T. 1992. *Reinventing government.* Reading, MA: Addison-Wesley.

Pfeiffer, L. W., Goodstein, L. D., and Nolan, T. M. 1989. *Shaping strategic planning: frogs, dragons, bees, and turkey tails.* Glenview, IL: Scott, Foresman.

Planning, Development and Evaluation Staff, Extension Service. January 4, 1994. *Profile of Extension.* Washington, D.C.: United States Department of Agriculture. Available: Internet bhewitt@esusda.gov

Popcorn, F. 1991. *The popcorn report.* New York, NY: Brain Reserves, Inc.

Rasmussen, W. D. 1989. *Taking the university to the people: seventy-five years of Cooperative Extension.* Ames, IA: Iowa State University Press.

Ratchford, C. B. Sept./Oct. 1984. Extension: unchanging, but changing. *Journal of Extension* 31:8–15.

Rockwell, S. K., Furgason, J., Jacobson, C., Schmidt, D., and Tooker, L. Fall 1993. From single to multicounty programming units. *Journal of Extension* 31(3): 17–19.

Segal, J. 1991. Basic needs, income and development. In *Ethics in agriculture: an anthology on current issues in world context,* C. V. Blatz, ed. Moscow, ID: University of Idaho Press.

Smith, K. L. Winter 1991. Philosophy diversions: which road? *Journal of Extension* 29:24–25.

Stevens, S. C. Spring 1991. Enhanced media use. *Journal of Extension* 29(1):37–38.

Toffler, E. 1970. *Future shock.* New York, NY: Bantam.

Toffler, E. 1990. *Power shift.* New York, NY: Bantam.

Tompkins, R. B. Winter 1989. No Time to be Timid. *Journal of Extension.* Vol. 27, 4–6.

United Way of America Strategic Institute. 1989. *Nine forces reshaping America.* Bethesda, MD: United Way.

U.S. Census. 1990. Washington, D.C.: Census Bureau.

Yates, A. C. 1994. *The renaissance of outreach in the land-grant tradition.* Seaman A. Knapp Memorial Lecture at the Annual Meeting of the National Association of State Universities and Land-Grant Colleges Board of Agriculture Plenary Session, Chicago, November, 1994. Available: e-mail almanac@esusda.gov send usda-speeches knapp 1107.

Warner, P. D., and Christenson, J. A. 1984. *Cooperative Extension: a nation-wide assessment.* Boulder, CO: Westview Press.

Warner, P. D., Rennekamp, R., and Nall, M. January 1996. *Structure and function of the Cooperative Extension Service.* (draft). Report submitted to Personnel and Organization Development–Extension Committee on Organizational Policy.

Glossary

Adoption—"A decision to make full use of an innovation as the best course of action." Rogers, Everett M. 1995. *Diffusion of Innovations* 4th ed., p. 21. New York, NY: The Free Press.

Agricultural programs—One of the program areas of Extension work. The main objective of the program is to help producers retain their competitiveness in world markets. The major goal is to increase profitability, create new and alternative opportunities, and use sound management practices that will help to preserve renewable natural resources.

Agricultural societies—Groups organized on a state, county, and community basis for the purpose of educating members about agricultural concerns, and for the promotion of agriculture in general. The first society devoted to agriculture was organized in Philadelphia, Pennsylvania in 1785. The societies served as a focal point of agitation for a department of agriculture in the federal government and the establishment of colleges to teach agriculture and the mechanic arts.

Base programs—Programs that address problems and issues that are common to most Extension units and central to the mission of the various Extension discipline areas. They may also be multi-disciplinary in nature. Base programs are on-going from year to year.

Bennett's Hierarchy—A model that has been used in Extension extensively to describe different levels of objectives and program accomplishments. Each of the seven levels of the hierarchy indicate a criteria for developing program objectives and for measuring program results.

Client—Individuals who Extension serves through the educational process. May include 4-H and other youth, adult volunteer leaders, and adult learners. Anyone who plans and participates in educational programs conducted by Extension.

Clubs—A group of youth or adults organized for a common purpose. Most clubs have officers and a program of work or activities to accomplish its mission. A club may be organized on a community basis (e.g., community or school boundary, or section of a city) or it may be organized to study specific interests, such as a photography project. Extension educators work primarily with 4-H clubs and Family Community Education clubs (formerly called Extension Homemaker clubs). Working through clubs is an efficient method for reaching and teaching specific audiences.

Coalition—"Individuals or organizations working together in a common effort for a common purpose to make more effective and efficient use of resources, an alliance." (Clark, Richard, et. al. 1995). *Building Coalitions.* p. 1. Columbus, OH: The Center for Action on Coalition Development, The Ohio State University Cooperative Extension Service.

Committee—A group of individuals who have the responsibility to perform a function, such as investigating, planning, or acting on a particular matter within a short period of time. Committees are used extensively in Extension work. Some examples include activity planning committees, program review committees, advisory committees, standing committees, and ad hoc committees.

Community Development Program—A program area of Extension work with the objective of improving the physical, social, and economic conditions of a community. The program goals are to strengthen communities by increasing group effectiveness, and to improve the quality and level of living that is public in nature. Extension professionals target citizen groups, business and industry leaders, city and county governing bodies, and voluntary service and civic organizations.

Cooperative Extension Service (CES)—A public-funded, nonformal, educational system that links the education and research resources of the U.S. Department of Agriculture (USDA), land-grant universities, and county administrative units. The basic mission of this system is to help people improve their lives through an educational process that uses scientific knowledge focused on issues and needs.

Cooperative farm-demonstration work—A program established by the federal government around 1887 where "agents" were appointed to work with farmers in small geographic areas to demonstrate new or improved production practices such as ways of controlling certain diseases of crops. The success of this demonstration method led to the employment of the first county agents.

Cooperative State Research, Education, Extension Service (CSREES)—National level of USDA organization for research and Extension established in 1994 under the Government Reorganization Act. Formerly Extension was a separate unit within USDA referred to as Extension Service—United States Department of Agriculture (ES-USDA).

County agent—The extension educator employed at the local county or parish level. The number of agents per county vary according to community size and support. The agent's primary responsibilities are educator and advisor, and transferring the findings of research and new technology to the solution of problems in the community, farm/ranch, or home. Specific title of this position may vary from-state to state. Some titles include: Farm Agent, Agriculture, Home Economics, or 4-H Agent, Youth Development Agent, Family and Consumer Science Educator.

Demonstration work (Also see Cooperative farm-demonstration work)—A demonstration of a new production practice, the test of a research recommendation or the trial of a new crop variety on a local producer farm. These methods include educational activities such as short courses, special interest meetings, clubs and community forums.

Design and Implementation—The step in the program planning process that builds on planning. Includes the selection and development of program content, selection and/or development of program delivery methods and resource materials, and creation of time lines for program implementation and evaluation. Also includes putting a program into operation.

Diffusion—"The process by which an innovation is communicated through certain channels over time among the members of a social system. Diffusion is a special type of communication concerned with the spread of messages that are new ideas." Rogers, Everett M. 1995. *Diffusion of Innovations* 4th ed., p. 5. New York: The Free Press.

Educational activity—Part of a program that has been planned and conducted to meet the stated objectives. Nonformal (non-credit) examples may include: meetings, workshops, discussion, field days, and medial presentations.

Epsilon Sigma Phi (ESP)—Epsilon Sigma Phi is an educational fraternity for Extension professionals who have exhibited excellence in education and programming leadership. The mission of ESP is dedicated to fostering standards of excellence in the Extension system and developing the Extension profession and the provisional.

Evaluation—The systematic process of determining the worth of a person, product, or program. Includes the planning and implementation procedures to measure various dimensions of program success and impact. Provides information for further action.

Expanded Food & Nutrition Education Program (EFNEP)—EFNEP is an integral part of the home economics and 4-H youth programs in Extension. It is designed to reach low-income families with young children. The goal of the program is to help families learn to improve dietary practices and become effective managers of available resources.

Extension Committee on Organizational Policy (ECOP)—A fourteen person committee, comprised of Extension directors from each of the four regions of the US. ECOP concerns itself with all aspects of the Extension organization: structure, programs, finances and relationships with others including USDA, research, resident instruction, and other organizations. All program and policy decisions related to Extension are addressed by ECOP.

Extension educators—Professional employees of the state Extension service of the land-grant institutions and the Extension Service-

USDA. May include county staff (agents, program assistants, EFNEP educators), District Staff (agents, directors, program specialists) and State staff (administrators, program specialists,).

Extension educational process—The composite of actions where an Extension educator conducts a situational analysis of individual and community needs, establishes specific learner objectives, implements a plan of work and evaluates the outcomes of the instruction to determine of behavioral changes have occurred.

Extension Management Information System (EMIS)—A reporting and accountability system that provides data on clientele contacts, clientele characteristics, time devoted to major programs or initiatives.

Extension partnership—The tripartite organizational structure of the Cooperative Extension System. Includes the federal partner (CSREES-Cooperative States Research, Education, and Extension Service - USDA, former ES-USDA), state partners (Extension services of the state land-grant universities), and local partners (county or parish legislative units).

Extension teaching methods—A collective phrase for all the known methods by which extension educators teach youth and adults. These methods are classified by the three functions of teaching: information delivery, program delivery and problem-solving tasks or by the method of contact: individual, group, and mass media methods.

Extension work—A collective phrase for describing the various methods by which extension educators accomplish the educational mission of the organization and the program areas that are central to its instruction.

Family Community Education Clubs (formerly Extension Homemaker clubs)—A volunteer organization at the community and county level to learn about subjects related to the family and the home. Local clubs have educational and community improvement activities. A county, state and national organizational structure assists members to develop leadership skills.

Farmers' Institutes—A community meeting covering a period of two to three days devoted to a discussion of agricultural problems and subjects related to the home. Many of these institutes were organized by state boards of agriculture as early as 1850 as a primary means of disseminating research findings of the Experiment Stations to those who could not or would not attend college.

Focus groups—An evaluation technique that utilizes small groups of individuals with a common interest who discuss a topic in a systematic manner.

Formative evaluation—Evaluations for reasons of improvement. Provides feedback throughout the program planning and implementation process allowing for adjustments or mid-course corrections.

4-H program—One of the program areas of Extension work with the objective of helping youth acquire the life skills and knowledge necessary to grow and succeed in a rapidly changing and complex society. The mission of the 4-H and Youth Development program is to create supportive environments in which culturally diverse youth and adults can reach their fullest potential.

4-H project—A structured learning experience for 4-H youth that enables them to learn, make, or do something. Project work can include both individual and group efforts that emphasize the "learn-by-doing" approach and incorporate real-life experiences. Project topics are available in a wide variety of interests to attract youth from all backgrounds, age, and levels of ability.

Full Time Equivalent (FTE)—One FTE equals one Extension employee working 228 work days.

Futuring—A planning strategy that encourages participants to think about the future; improve understanding of trends and their implications; anticipate consequences of intended or unintended behaviors, decisions, and policies; and enhance vision of a desirable future. Futuring is a technique used for long-range or strategic planning. Common approaches include: forecasting, simulations, risk management, and computer modeling.

Goals—A clearly written statement about the desired direction or outcome of an educational program. May be determined by an individual or group.

Group teaching methods—A broad category of teaching for a number of persons assembled together in one location. These methods include such educational activities as short courses, special interest meetings, clubs, and community forums.

Hatch Act of 1887—An act that provided for the establishment of an agricultural experiment station in each of the land-grant colleges established under the Morrill Act of 1862. This act established research as one of the functions of the land-grant colleges and universities.

Home economics programs—A program area of Extension work with emphasis on teaching family members how to improve the social, economic, and physical well-being of their families. Its goals are to assist families to manage resources better, make sound decisions, improve the level of nutrition, diet and health, and build human capital.

Individual contact teaching methods—One-on-one instructional methods used by Extension educators for imparting knowledge to clientele. Some individual methods include farm, home, or office visits.

Innovation—"An idea, practice, or object that is perceived as new by an individual or other unit of adoption." Rogers, Everett M. 1995. *Diffusion of Innovations* 4th ed., p. 11. New York, NY: The Free Press.

International Four-H youth Exchange (IFYE)—A two-way exchange of United States 4-H members and youth from other countries throughout the world. Young people ages 19 to 28 live with rural families in another country.

Issues—Matters of wide public concern arising out of complex human problems.

Issues programming—Educational efforts that focus on issues facing individuals in their own context or setting. Issues programming involves identifying the problem or issue, the audience, the method of program implementation, and the resources necessary to address the issue.

Knapp, Seaman A.—An educator, farmer, and land developer. He was president of the Iowa Agricultural College and in charge of the USDA farm demonstration work. Knapp is called the father of Extension work. He is largely responsible for the success of the cooperative farm-demonstration work and the development of boys and girls clubs in the South. This method became the model for federal Extension efforts and the primary method of teaching outlined in the Smith-Level Act.

Land-grant college/university—"An institution of higher education sustained and supported by the Morill Acts of 1862 and 1890, and expanded by the Hatch Act of 1887, the Smith-Lever Act of 1914, and subsequent legislation." Sanderson, David R. 1988. *Working with Our Publics: Module 1.* p. 59. Raleigh, NC: North Carolina Agricultural Extension Service, North Carolina State University.

Land-grant philosophy—Education outreach with emphasis on practical and useful knowledge, linkage to research, the use of hands-on approach, and programming in a nonformal or non-school setting.

Leadership—Influence, or the art or process of influencing people, so that they strive willingly and enthusiastically toward the accomplishment of group goals.

Management—Effective utilization of available resources (human and material) to achieve common goals or objectives and has been described as an art, a science, a body of knowledge, a process, and a set of functions.

Mass-media teaching methods—A broad category of teaching methods that utilizes print or broadcast media. Some of these media can include newspaper articles, television programs, or satellite broadcasts.

Memorandum of Understanding—This memorandum outlines the responsibilities and clarifies the working relationship of the partners involved in implementing the Smith-Lever Act.

Morrill Act of 1862—An act for the endowment, support, and maintenance of at least one college in a state whose primary objective was to teach agriculture and the mechanic arts without excluding other scientific and classical studies.

Morrill Act of 1890—A second land-grant act that provided for the maintenance of the land-grant colleges through endowment funds, sometimes called the 2nd Morrill Act. It specified no federal support for colleges that made a distinction of admission regarding race or color. It allowed for those existing state colleges for white and black students to receive funds, provided the funds were distributed equitably and for the creation of separate, but equal facilities for new colleges. Some seventeen "historically black" institutions are known as 1890 land-grant colleges.

National 4-H Council—The National 4-H Council is a private, not-for-profit educational partner of 4-H and Extension. The Council seeks major grants and donations to support educational activities and programs of youth, adults, and volunteers involved in Extension and 4-H work. The council also operates the National 4-H Conference Center in Chevy Chase, Maryland, the National 4-H Supply Service, and a national marketing firm for 4-H. The National 4-H Council mission is to build partnerships for community youth development that value and involve youth in solving issues critical to their lives, their families, and society.

National initiatives—The focus of Extension work toward the most current, significant, and complex issues that the organization has the potential to impact. These issues are widespread across the various states and are identified at the federal level.

Needs assessment—The systematic process of analyzing gaps between what learners know and can do, and what they should know and be able to do.

Nonformal education—"Out-of-school, noncredit education formats; the essential form of Extension education." Sanderson, David R. 1988. *Working with Our Publics: Module 1.* p. 59. Raleigh, NC: North Carolina Agricultural Extension Service, North Carolina State University.

Objectives—Precise statements defining what program participants should be able to do upon completion of the program. Objectives state the intended changes in individuals, groups, or communities as a result of the educational program.

Organic Act of 1862—An act that created the United States Department of Agriculture.

Orientation—Familiarizing staff and volunteers with both the total organization and with specific job responsibilities.

Paraprofessional—Supervised by and serves as an assistant to the county agent. Also known as program assistant. Most frequently assists with coordinating and organizing many of the activities and events that the office sponsors.

Plan of work (POW)—Plan of action that puts the educational programs in operation. Involves setting specific objectives in terms of the behavioral change desired, the selection of teaching methods, the scheduling of activities, and the dividing of responsibilities among Extension professionals for each program area. Developed at all levels (individual, county, and state). Individual and county plans are reviewed annually. State plans are developed and submitted to USDA on a four-year budget cycle. All POWs are based upon program priorities determined from Extension needs assessment throughout the state. Generally includes several elements: a statement of the situation of need; program objectives; identification of target audiences; program delivery methods; and strategies for evaluation.

Professional Research and Knowledge (PRK)—A 4-H data base established to catalog the knowledge and research base of 4-H youth development programs. PRK, introduced in 1986 and revised in 1994, also includes a taxonomy of competencies and skills needed by 4-H professionals.

Program (Extension education program)—A planned sequence of educational experiences guided by specific objectives. Includes activities and events that are planned, conducted, and evaluated for their impact on identified needs of the participants. Usually occurs over a period of time.

Program development—The Extension Committee on Policy (1974) defined Extension program development as a continuous series of complex, interrelated processes which result in the accomplishments of the educational mission and objectives of the organization. Most Extension planning models identify three main components of the program development process: planning, design and implementation, and evaluation and accountability. The terms program planning and program development are frequently used interchangeably.

Program planning—See program development.

Situational analysis—A description of the community and circumstances surrounding a proposed program that enables the educator to understand the environment for programming more completely. May include social, historical, educational, emotional, economic, political, and personal factors.

Smith-Lever Act of 1914—An act that provided for mutual cooperation of USDA and land-grant colleges in conducting agricultural Extension work.

Social indicators—Demographic or statistical data that describe the size and characteristics of population groups. Alone, these indicators may not establish need, but in combination with other information, they can provide evidence of need.

Specialists—Faculty members with expertise and specialized knowledge in a particular subject-matter area. Involved in translating and disseminating researched-based material to county Extension agents and their clientele groups. Specialists usually have a doctorate degree with rank equivalent to the campus professor system.

Staff development and training—Assisting staff and volunteers to develop additional knowledge, attitudes, and skills that will help them do their jobs better and more responsibly.

Stakeholders—People who have a vested interest in a program.

Summative Evaluation—Accountability evaluations provides documentation of program outcomes and end results. This information is critical in documenting program impact, making changes for future programs, and identifying additional goals and objectives for future programming.

Supervision—The process of leading and influencing individuals and groups to assist willingly and harmoniously in accomplishing objectives established by the organization. To ensure accomplishment of objectives, the supervisor is involved in establishing and communicating standards, measuring performance against those standards, and acknowledging success and reviewing areas of concerns.

Support staff—Staff may include receptionists, secretaries, bookkeepers, and office managers. They have a vital function in any organization and are often the ones who have the initial contact with clients.

Target audiences—A group of individuals with special needs who become the focus of Extension teaching.

United States Department of Agriculture (USDA)—A department of the federal government established by President Lincoln in 1862 under the Organic Act for the purpose of acquiring and diffusing useful information on subjects related to agriculture and to distribute among the people new and valuable seeds and plants. The first Secretary of Agriculture (a cabinet position) was named in 1889.

Volunteers—Unpaid individuals who offer their time, talents, and services to the organization.

Appendix 1:

Smith-Lever Act
May 8, 1914

Be it enacted by the Senate and House of Representatives of the United States of America in Congress assembled, That in order to aid in diffusing among the people of the United States useful and practical information on subjects relating to agriculture and home economics, and to encourage the application of the same, there may be inaugurated in connection with the college or colleges in each State now receiving, or which may hereafter receive, the benefits of the Act of Congress approved July second, eighteen hundred and sixty-two, entitled "An Act donating public lands to the several States and Territories which may provide colleges for the benefit of agriculture and the mechanic arts" (Twelfth Statues at Large, page five hundred and three), and of the Act of Congress approved August thirtieth, eighteen hundred and ninety (Twenty-sixth Statutes Large, page four hundred and seventeen and chapter eight hundred and forty-one), agricultural extension work which shall be carried on in cooperation with the United States Department of Agriculture: *Provided*, That in any State in which two or more such colleges have been or hereafter may be established, the appropriations hereinafter made to such State shall be administered by such college or colleges as the legislature of such State may direct: *Provided further*, That, pending the inauguration and development of the cooperative extension work herein authorized, nothing in this Act shall be construed to discontinue either the farm management work or the farmers' cooperative demonstration work as now conducted by the Bureau of Plant Industry of the Department of Agriculture.

SEC. 2. That cooperative agricultural extension work shall consist of the giving of instruction and practical demonstrations in agriculture and home economics to persons not attending or resident in said colleges in the several communities, and imparting to such persons information on said subjects through field demonstrations, publications, and otherwise; and this work shall be carried on in such manner as may be mutually agreed upon by the Secretary of Agriculture and the State agricultural college or colleges receiving the benefits of this Act.

SEC. 3. That for the purpose of paying the expenses of said cooperative agricultural extension work and the necessary printing and distributing of information in connection with the same, there is permanently appropriated, out of any money in the Treasury not otherwise appropriated, the sum of $480,000 for each year, $10,000 of which shall be paid annually, in the manner hereinafter provided, to each State which shall by action of its legislature assent to the provisions of this Act: *Provided*, That payment of such installments of the appropriation herein before made as shall become due to any State before the adjournment of the regular session of the legislature meeting next after the passage of this Act may, in the absence of prior legislative assent, be made upon the assent of the governor thereof, duly certified to the Secretary of the Treasury: *Provided further,* That there is also appropriated an additional sum of $600,000 for the fiscal year following that in which the foregoing appropriation first becomes available, and for each year thereafter for seven years a sum exceeding by $500,000 the sum appropriated for each preceding year, and for each year thereafter there is permanently appropriated for each year the sum of $4,100,000 in addition to the sum of $480,000 hereinbefore provided: *Provided further,* That before the funds herein appropriated shall become available to any college for any fiscal year plans for the work to be carried on under this Act shall be submitted by the proper officials of each college and approved by the Secretary of Agriculture. Such additional sums shall be used only for the purposes hereinbefore stated, and shall be allotted annually to each State by the Secretary of Agriculture and paid in the manner hereinbefore provided, in the proportion which the rural population of each State bears to the total rural population of all the States as determined by the next preceding Federal census: *Provided further,* That no payment out of the additional appropriations herein provided shall be made in any year to any State until an equal sum has been appropriated for that year by the legislature of such State, or provided by State, county, college, local authority, or individual contributions from within the State, for the maintenance of the cooperative agricultural extension work provided for in this Act.

SEC. 4. That the sums hereby appropriated for extension work shall be paid in equal semiannual payments on the first day of January and July of each year by the Secretary of the Treasury upon the warrant of the Secretary of Agriculture, out of the Treasury of the United States, to the treasurer or other officer of the State duly authorized by the laws of the State to receive the same; and such officer shall be required to report to the Secretary of Agriculture, on or before the first day of September of each year, a detailed statement of the amount so received during the previous fiscal year, and of its disbursement, on forms prescribed by the Secretary of Agriculture.

SEC. 5. That if any portion of the moneys received by the designated officer of any State for the support and maintenance of cooperative agricultural extension work, as provided in this Act, shall by any action or contingency be diminished or lost, or be misapplied, it shall be replaced by said State to which it belongs, and until so replaced no subsequent appropriation shall be apportioned or paid to said State, and no portion of said moneys shall be applied, directly or indirectly, to the purchase, erection, preservation, or repair of any building or buildings, or the purchase or rental of land, or in college-course teaching, lectures in colleges, promoting agricultural trains, or any other purpose not specified in this Act, and not more than five per centum of each annual appropriation shall be applied to the printing and distribution of publications. It shall be the duty of each of said colleges annually, on or before the first day of January, to make to the governor of the State in which it is located a full and detailed report of its operations in the direction of extension work as defined in this Act, including a detailed statement of receipts and expenditures from all sources for this purpose, a copy of which report shall be sent to the Secretary of Agriculture and to the Secretary of the Treasury of the United States.

SEC. 6. That on or before the first day of July in each year after the passage of this Act the Secretary of Agriculture shall ascertain and certify to the Secretary of the Treasury as to each State whether it is entitled to receive its share of the annual appropriation for cooperative agricultural extension work under this Act, and the amount which it is entitled to receive. If the Secretary of Agriculture shall withhold a certificate from any State of its appropriation, the facts and reasons therefor shall be reported to the President, and the amount involved shall be kept separate in the Treasury until the expiration of the Congress next succeeding a session of the legislature of any State from which a certificate has been withheld, in order that the State may, if it should so desire, appeal to Congress from the determination of the Secretary of Agriculture. If the next Congress shall not direct such sum to be paid, it shall be covered into the Treasury.

SEC. 7. That the Secretary of Agriculture shall make an annual report to Congress of the receipts, expenditures, and results of the cooperative agricultural extension work in all of the States receiving the benefits of this Act, and also whether the appropriation of any State has been withheld; and if so, the reasons therefor.

SEC. 8. That Congress may at any time alter, amend, or repeal any or all of the provisions of this Act.

Approved, May 8, 1914.

Appendix 2:

Smith-Lever Act (Amended November 28, 1990)

August 1991—As amended by the Food, Agriculture, Conservation, and Trade Act of 1990, P.L. 101-624, 11/28/90.

SEC. 1. In order to aid in diffusing among the people of the United States useful and practical information on subjects relating to agriculture and home economics, and rural energy, and to encourage the application of the same, there may be continued or inaugurated in connection with the college of the or colleges in each State, Territory, or possession, now receiving, or which may hereafter receive, the benefits of the Act of Congress approved July second, eighteen hundred and sixty-two, entitled "An Act donating public lands to the several States and Territories which may provide colleges for the benefit of agriculture and the mechanic arts," and of the Act of Congress approved August thirtieth, eighteen hundred and ninety, agricultural extension work which shall be carried on in cooperation with the United States Department of Agriculture. *Provided,* That in any State, Territory, or possession in which two or more such colleges have been or hereafter may be established, the appropriations hereinafter made to such State, Territory, or possession shall be administered by such college or colleges as the legislature of such State, Territory, or possession may direct.

SEC. 2. Cooperative agricultural extension work shall consist of the development of practical applications of research knowledge and giving of instruction and practical demonstrations of existing or improved practices or technologies in agriculture, home economics, and rural energy, and subjects relating thereto to persons not attending or resident in said colleges in the several communities, and imparting information on said subjects through demonstrations, publications, and otherwise and for the necessary printing and distribution of information in connection with the foregoing: and this work shall be carried on in such manner as may be mutually agreed upon by the Secretary of Agriculture and the State agricultural college or colleges or Territory possession receiving the benefits of this Act.

(a) There are hereby authorized to be appropriated for the purposes of this Act such sums as Congress may from time to time determine to be necessary.

(b)(1) Out of such sums, each State and the Federal Extension Service shall be entitled to receive annually a sum of money equal to the sums available from Federal cooperative extension funds for the fiscal year 1962, and subject to the same requirements as to furnishing of equivalent sums by the State, except that amounts heretofore made available to the Secretary for allotment on the basis of special needs shall continue available for use on the same basis.

(b)(2) There is authorized to be appropriated for the fiscal year ending June 30, 1971, and for each fiscal year thereafter, for payment to the Virgin Islands, Guam, and the Northern Mariana Islands, $100,000 each, which sums shall be in addition to the sums appropriated for the several States of the United States and Puerto Rico under the provisions of this section. The amount paid by the Federal Government to the Virgin Islands and Guam pursuant to this paragraph shall not exceed during any fiscal year, except the fiscal years ending June 30, 1971, and June 30, 1972, when such amount may be used to pay the total cost of providing services pursuant to this Act, the amount available and budgeted for expenditure by the Virgin Islands and Guam for the purposes of this Act.

(c) Any sums made available by the Congress for further development of cooperative extension work in addition to those referred to in subsection (b) hereof shall be distributed as follows:

1. Four per centum of the sum so appropriated for each fiscal year shall be allotted to the Federal Extension Service for administrative, technical, and other services, and for coordinating the extension work of the Department and the several States, Territories, and possessions.

2. Of the remainder so appropriated for each fiscal year, 20 per centum shall be paid to the several States in equal proportions, 40 per centum shall be paid to the several States in the proportion that the rural population of each bears to the total rural population of the several States as determined by the census, and the balance shall be paid to the several States in the proportion that the farm population of each bears to the total farm population of the several States as determined by the census: *Provided*, That payments out of the additional appropriations for further development of extension work authorized herein may be made subject to the making available of such sums of public funds by the States from non-Federal funds for the maintenance of cooperative agricultural extension work provided for in this Act, as may be provided by the Congress at the time such additional appropriations are made: *Provided further*, That any appropriation made hereunder shall be allotted in the first and succeeding years on the basis of the decennial census current at the time such appropriation is first made, and as to any increase, on the basis of decennial census current at the time such increase is first appropriated.

(d) The Federal Extension Service shall receive such additional amounts as Congress shall determine for administration, technical, and other services and for coordinating the extension work of the Department and the several States, Territories, and possessions.

(e) Insofar as the provisions of subsections (b) and (c) of this section, which require or permit Congress to require matching of Federal funds, apply to the Virgin Islands of the United States and Guam, such provisions shall be deemed to have been satisfied, for the fiscal years ending September 30, 1978, and September 30, 1979, only, if the amounts budgeted and available for expenditure by the Virgin Islands of the United States and Guam in such years equal the amounts budgeted and available for expenditure by the Virgin Islands of the United States and Guam in the fiscal year ending September 30, 1977.

(f)(1) The Secretary of Agriculture may conduct educational, instructional, demonstration, and publication distribution programs through the Federal Extension Service and enter into cooperative agreements with private nonprofit and profit organizations and individuals to share the cost of such programs through contributions from private sources as provided in this subsection.

(f)(2) The secretary may receive contributions under this subsection from private sources for the purposes described in paragraph (1) and provide matching funds in an amount not greater than 50 percent of such contributions.

SEC. 4. On or about the first day of July in each year after the passage of this Act, the Secretary of Agriculture shall ascertain as to each State whether it is entitled to receive its share of the annual appropriation for cooperative agricultural extension work under this Act and the amount which it is entitled to receive. Before the funds herein provided shall become available to any college for any fiscal year, plans for the work to be carried on under this Act shall be submitted by the proper officials of each college and approved by the Secretary of Agriculture. The Secretary shall ensure that each college seeking to receive funds under this Act has in actual or potential conflicts of interest among employees of such colleges whose salaries are funded in whole or in part with such funds. Such sums shall be paid in equal quarterly payments in or about October, January, April, and July of each year to the treasurer or other officer of the State duly authorized by the laws of the State to receive the same, and such office shall be required to report to the Secretary of Agriculture on or about the first day of April of each year, a detailed statement of the amount so received during the previous fiscal year and its disbursement, on forms prescribed by the Secretary of Agriculture.

SEC. 5. If any portion of the moneys received by the designated officer of any State for the support and maintenance of cooperative agricultural extension work, as provided in this Act, shall by any action or contingency be diminished or lost or be misapplied, it shall be replaced by said State and until so replaced no subsequent appropriation shall be apportioned or paid to said State. No portion of said moneys shall be applied, directly or indirectly, to the purchase, erection, preservation, or repair of any building or buildings, or the purchase or rental of land, or in college-course teaching, lectures in college, or any other purpose not specified in this Act. It shall be the duty of said colleges, annually, on or about the first day of January, to make the Governor of the State in which it is located a full and detailed report of its operations in extension work as defined in this Act, including a detailed statement of receipts and expenditures from all sources for this purpose, a copy of which report shall be sent to the Secretary of Agriculture.

SEC. 6. If the Secretary of Agriculture finds that a State is not entitled to receive its share of the annual appropriation, the facts and reasons therefor shall be reported to the President, and the amount involved shall be kept separate in the Treasury until the expiration of the Congress next succeeding a session of the legislature of the State from which funds have been withheld in order that the State may, if it should so desire, appeal to Congress from the determination of the Secretary of Agriculture. If the next Congress shall not direct such sum to be paid, it shall be covered into the Treasury.

SEC. 7. Repealed. (Dealt with an annual report to Congress.)

SEC. 8. (a) The Congress finds that there exists special circumstances in certain agricultural areas which cause such areas to be at a disadvantage insofar as agricultural development is concerned, which circumstances include the following:

(1) Therefore is concentration of farm families on farms either too small or too unproductive or both;

(2) such farm operators because of limited productivity are unable to make adjustments and investments required to establish profitable operations;

(3) the productive capacity of the existing farm unit does not permit profitable employment of available labor;

(4) because of limited resources, many of these farm families are not able to make full use of current extension programs designed for families operating economic units nor are extension facilities adequate to provide the assistance needed to produce desirable results.

(b) In order to further the purposes of section 2 in such areas and to encourage complementary development essential to the welfare of such areas, there are hereby authorized to be appropriated such sums as the Congress from time to time shall determine to be necessary for payments to the States on the basis of special needs in such areas as determined by the Secretary of Agriculture.

(c) In determining that the area has such special need, the Secretary shall find that it has substantial number of disadvantaged farms or farm families for one or more of the reasons heretofore enumerated. The Secretary shall make provisions for the assistance to be extended to include one or more of the following:

(1) Intensive on-the-farm educational assistance to the farm family in appraising and resolving its problem;

(2) assistance and counseling to local groups in appraising resources for capability of improvements in agriculture or introduction of industry designed to supplement farm income;

(3) cooperation with other agencies and groups in furnishing all possible information as to existing employment opportunities, particularly to farm families having underemployed workers; and

(4) in cases where farm family, after analysis of its opportunities and existing resources, finds it advisable to seek a new farming

venture, the providing of information, advice, and councel in connection with making such change.

(d) No more than 10 per centum of the sums available under this section shall be allotted to any one State. The Secretary shall use project proposals and plans of work submitted by the State Extension directors as a basis for determining the allocation of funds appropriated pursuant to this section.

(e) Sums appropriated pursuant to this section shall be in addition to, and not in substitution for, appropriations otherwise available under this Act. The amounts authorized to be appropriated pursuant to this section shall not exceed a sum in any year equal to 10 per centum of sums otherwise appropriated pursuant to this Act.

SEC. 9. The Secretary of Agriculture is authorized to make such rules and regulations as may be necessary for carrying out the provisions of this Act.

SEC. 10.[1] The term "State" means the States of the Union and Puerto Rico, the Virgin Islands, Guam, and the Northern Mariana Islands. (Code reference is 7 U.S.C. 341 et. seq.)

[1]P.L. 96-374, Section 1361(c) states: Any provision of any Act of Congress relating to the operation or provision of assistance to a land-grant college in the Virgin Islands or Guam shall apply to the land-grant college in American Samoa and in Micronesia in the same manner and to the same extent.

Appendix 3:

Chronological History of Legislation by the United States Congress Relating to the Cooperative Extension Service in the Land-Grant Universities

Year	Title	Purpose
1862	Morrill Act (first)	Donating public lands to the several states and territories which may provide colleges for the benefit of agriculture and the mechanic arts.
1887	Hatch Act	To establish agricultural experiment stations in connection with the colleges in the several states operating under the provisions of the 1862 Morrill Act with the provision that each state provide funds matching those of the Federal government.
1890	Morrill Act (second)	To apply a portion of the proceeds of the public lands to the more complete endowment and support of the colleges for the benefit of agriculture and the mechanic arts established under the first Morrill Act and the establishment of such colleges for the black race in those states prohibiting their attendance at white institutions.
1914	Smith-Lever Act	To provide for cooperative agriculture extension work between the agricultural colleges in the several states receiving the benefits of the 1862 and 1890 land-grant acts.
1924	Clark-McNary Act	Section 5 of the Act provided funds (on a matching basis by the individual states) for cooperative farm-forestry work.
1928	Hawaii Act	An act extending the benefits of the Hatch Act and Smith-Lever Act to the Territory of Hawaii.
1928	Capper-Ketcham Act	To provide for the further development of agricultural extension work at the 1862 land-grant colleges and that future funds be allocated "in addition to and not a substitute for" those made available in the Smith-Lever Act of 1914.
1929	Alaska Act of 1929	To extend the benefits of the Hatch Act and the Smith-Lever Act to the Territory of Alaska.
1931	Puerto Rico Act	To coordinate the agricultural-experiment station work and to extend the benefits of the Hatch and Smith-Lever Act to the Territory of Puerto Rico.
1935	Bankhead-Jones Act	An act extending the scope of research conducted under the Hatch Act and to provide for the future development of Cooperative Agricultural Extension work and to provide for

		the further endowment and support of 1862 and 1890 land-grant colleges.
1936	Alaska Act of 1936	An act extending benefits of the Capper-Ketcham Act to the Territory of Alaska.
1937	Puerto Rico Bankhead-Jones Act	An act extending the benefits of extension (Sec. 21) of the Bankhead-Jones Act to Puerto Rico.
1939	Act of 1939	An act to provide additional funding and further development of agricultural extension work being conducted under the Smith-Lever Act of 1914.
1945	Bankhead-Flannagan Act	To provide additional funding and further development of the cooperative agricultural extension work under the Smith-Lever Act of 1914 and Bankhead-Jones Act of 1935.
1946	Agricultural Marketing Act	Authorized extension programs in marketing, transportation, distribution of agricultural products outside Smith-Lever formula, but states required to match Federal funds.
1949	Clarke-McNary Amendment	Authorized USDA to cooperate with land-grant colleges in aiding farmers through advice, education, demonstration, etc., in establishing, renewing, protecting, and managing wood lots, etc., and in harvesting, utilizing, and marketing the products thereof.
1953	Smith-Lever Act Amendment	An act that simplified and consolidated ten separate laws relating to extension. Established new funding procedures based on rural/urban population formula and amounts. Repealed the Capper-Ketcham Act and the two Bankhead-Jones Acts of 1935 and 1945. Inserted "and subjects relating thereto" after agriculture and home economics and inserted reference to necessary printing and distribution of information.
1955	Smith-Lever Amendment	Authorized work with disadvantaged farms and farm families and authorized funds for extension outside the traditional funding "formula."
1962	Smith-Lever Amendment	Inserted "or territory or possession" after "college or colleges."
1966	National Sea Grant College and Program Act	Established a program (under the U.S. Department of Commerce) to provide for applied research, formal education, and advisory (extension) services for development of marine and Great Lakes resources. About two-thirds (of the 30 coastal and Great Lakes states involved) have integrated this effort with that of Cooperative Extension.

1968	District of Columbia Public Education Act	Designated Federal City College as the land-grant institution for extension in the District of Columbia and authorized funds for the work.
1972	Rural Development Act of 1972 - Title V	Authorized rural development and small-farm extension programs, required that administration of program be associated with program under Smith-Lever Act (Memorandum of Understanding required), and established State Rural Development Advisory Council.
1972	Smith-Lever Amendment	Virgin Islands and Guam designated as States under Section 10.
1977	Food and Agriculture Act	A very comprehensive act that affected Cooperative Extension in the following ways: Established a joint council on Food and Agricultural Sciences to foster coordination of the research, extension, and teaching activities in the Federal Government, the states, and among private and public colleges and universities. Established a National Agricultural Research and Extension Users Advisory Board. Authorized fixed amounts for extension activities through FY 1978 ($260,000,000) to FY 1982 ($350,000,000). Established a National Food and Human Nutrition Research and Education Program. Authorized agricultural and forestry extension funds for the 1890 institutions and Tuskegee Institute. Specifies that 4 percent of funds under Smith-Lever Act must go for extension work at these institutions and instructs extension heads at 1862 and 1890 institutions to develop a state-wide comprehensive plan. Amended the Rural Development Act of 1972 to provide additional assistance to small farmers (any farmer with gross sales of less than $20,000) and provides for use of paraprofessionals in this effort. Added the uses of solar energy with respect to agriculture; defined solar energy and allowed for establishment of solar energy project demonstrations. Required the Secretary to submit an evaluation of the Extension Service and the several Cooperative Extension Services by March, 1979.

		Directed the Secretary to assist Agency for International Development with agricultural research and extension programs in developing countries.
1978	Renewable Resources Extension Act	A very comprehensive act that affected Cooperative Extension in the following ways: Provided for educational programs concentrating on renewable resources, which includes fish and wildlife management, range management, timber management, and watershed management, as well as forest and range-based outdoor recreation opportunities, trees and forests in urban areas, and trees and shrubs in shelter belts. Required a five year plan for implementing the Act to be submitted to the National Agricultural Research and Extension Users Advisory Board.
1980	Smith-Lever Amendment	Inserted reference to rural energy in Section 2.
1981	Agriculture and Food Act of 1981	Amended the Food and Agriculture Act of 1977 as follows: Defined the terms "food and agricultural science," "state," "teaching," and "state cooperative agents," and cooperating forestry school. Added authority of the Secretary of Agriculture to coordinate all agricultural research, extension, and teaching activities conducted or financed by the Department of Agriculture, provided by the Forest and Rangeland Renewable Resources Planning Act of 1974, and by the Soil and Water Resources Conservation Act of 1977; and to develop long-range planning with all states, state institutions, Joint Council, Advisory Board, and other appropriate institutions. Defined the primary responsibility of the Joint Council; specified the terms of membership; and changed the dates for submitting the priority recommendations to the Secretary of Agriculture. Outlined the membership of the National Agricultural Research and Extension Users Advisory Board and changed the dates for submitting priority recommendations to the Secretary. Supported the Federal-State partnership.

		Provided for the employment and training of professionals and paraprofessional aides to engage in nutrition education of low-income families. Also set the appropriations and poverty guide lines for the expanded food and nutrition education program.
		Specified the appropriation for extension work of 1890 land-grant colleges, including Tuskegee Institute; changed title to "extension administrator" and required the submission of a plan of work.
		Authorized appropriations for solar energy model farms and demonstration projects.
		Gives the Secretary of Agriculture authority to coordinate research and extension efforts in international agriculture, which may include technical services, funds enhancement, or linkages among institutions.
		Authorized appropriations for extension programs.
		Authorized aquaculture extension work and rangeland research.
		Authorized rural development programs and small farm extension programs.
		Amended the Rural Development Act of 1972.
		Authorized the Secretary of Agriculture to conduct a regular evaluation of agricultural, research, extension, and teaching programs.
		Established Cooperative Research and Extension programs in aquatic food species.
		Established an Aquaculture Advisory Board.
1985	Food Security Act of 1985	Amended the National Agricultural Research Extension and Teaching Policy Act of 1977 to give authority to the Secretary of Agriculture to coordinate the efforts of states, extension services, the Joint Council, the Advisory Board, and others in developing a plan for the effective transfer of technology to the farming community, addressing the unique problems of small and medium sized farmers; established controls in the development and use of application of biotechnology to agriculture.
		Charged the National Agricultural Research and Extension Users Advisory Board with developing a plan for the effective transfer of new technologies to the farming community.
		Repealed Sections 1424 and 1427 of the 1977 Act regarding food and human nutrition research and extension program materials.

		Provided grants to upgrade 1890 land-grant college extension facilities, including Tuskegee Institute.
1985	Smith-Lever Amendment	Revised Section 2 to read: "Shall consist of the development of practical applications of research knowledge and giving of instruction and practical demonstrations of existing or improved practices or technologies." Amended by adding: The Secretary of Agriculture may conduct educational, instructional, demonstration, and publication distribution programs through the Federal Extension Service and enter into cooperative agreements with private nonprofit and profit organizations and individuals to share the cost of such programs through contributions from private sources as provided in this subsection. Authorized grants to 1890 Universities to improve Extension facilities and equipment. Authorized the Secretary of Agriculture to receive contributions from private sources and to provide matching funds not to exceed 50 percent of the contribution.
1988	Smith-Lever Amendment	Expanded the definition of "state" to include "States of the Union, the District of Columbia, Puerto Rico, the Virgin Islands, Guam, American Samoa, the Northern Mariana Islands, and the trust territory of the Pacific Islands."
1990	Conservation Program Improvement Act and the Food, Agriculture, Conservation and Trade Act of 1990	Directed the Extension Service to operate a program in each state to catalogue the federal, state, and local laws and regulations that govern the handling of unused or unwanted agricultural chemicals and agricultural chemical containers in the state. It also provided for the development of educational materials to be distributed to agricultural producers and the general public. Charged the Extension Service to have educational efforts that would inform the public of the safest way to compost and teach composing techniques and procedures for using compost. Expanded natural resource educational programs in the Renewable Resources Extension Act of 1978. Established a water quality coordination program which increased educational efforts regarding water contamination. Provided for assistance for the control of weeds and pests.

		Allowed for the expansion of the Expanded Food and Nutrition Education Program. Provided for the collaborative programs of food, nutrition, and consumer education for low-income individuals with other public agencies. Repealed Sections 3241-3271 regarding solar energy information system, farm and demonstration projects, and research and development centers. Established five regional aquaculture research and demonstration centers for research and Extension work and demonstration projects.
1990	The National Forest Dependent Rural Communities Economic Diversification Act of 1990	Expanded educational programs to rural communities that are economically dependent upon forest resources to assist in diversifying the community economic base. Specified training and education programs to be conducted by the Extension service.
1994	National Agricultural Research, Extension and Teaching Act of 1994	Established extension education programs on Native American reservations and tribal jurisdictions in states where located. Provided for technical assistance and training to Native American tribes and Alaskan Natives for subsistence farming programs. Provided grants for a pilot project to coordinate food and nutrition education programs of states. Provided for demonstration grants for extension and nonprofit disability agencies to provide on-the-farm agricultural education and assistance directed at accommodating disability in farm operations.
1994	The Department of Agricultural Reorganization Act of 1994	Established the Cooperative State Research, Education, and Extension Service to coordinate USDA and state cooperative agricultural research, extension, and education programs. Established the Cooperative State Research and Education Service to consolidate cooperative research and agricultural extension and education programs with state agricultural experiment stations and extension services within land-grant and related universities.

Mandated Extension Programs From Legislative Amendments

nutrition and family education
urban gardening
pest management
farm safety and rural health
pesticide impact assessment
rural development
groundwater quality
financially stressed and/or dislocated farmers
agriculture telecommunications
youth-at-risk
food safety
renewable resources
subsistence farming on Native American reservations
establish and operate centers of rural technology
outreach and assistance for socially disadvantaged farmers and ranchers
rural health and safety education
nutrition education and consumer education
1890 extension work
natural-resource-based economic development

Appendix 4:

A Profile of Extension's Professional Staff 1914 to 1994

Number of Extension Personnel by Type						
Year	Directors Administrators	State Specialists	Leaders and Supervisors	Area Agents*	County Agents	Total
1914	50	221	112	0	1,237	1,620
1918	115	512	575	0	5,526	6,728
1928	106	1,004	376	0	3,675	5,161
1938	131	1,551	493	0	6,507	8,682
1948	159	1,933	596	0	8,785	11,473
1958	217	2,554	754	0	11,124	14,649
1968	295	3,850	695	0	10,220	15,606
1978	487	3,410	696	732	11,342	16,667
1982	507	3,706	651	629	11,240	16,733
1986	601	4,322	602	619	10,375	16,519
1989	786	4,725	538	616	9,703	16,368
1994	896	5,169	609	474	9,760	16,908

*The category of Area Agent was not used prior to 1969.

Source: The table is adapted and updated from Warner and Christenson (1984, p. 13). Source of Data: Salary Analyses of Cooperative Extension Service Positions, December 1994: State Research, Education and Extension Service, USDA, Data for 1914–1994: U.S. Department of Agriculture, 1980a:30 and explanatory notes, 1984 Budget, 1983. Data for 1986: Office of Personnel Management, ES-USDA.

Appendix 5:

The 105 Land-Grant Colleges and Universities

March 1995

***Alabama**
Alabama A&M University* (Normal, AL)
Auburn University (Auburn, AL)
Tuskegee University (Tuskegee, AL)

Alaska
University of Alaska Statewide System (Fairbanks, AK)

American Samoa
Community College of American Samoa (Pago Pago, AQ)

Arizona
Navajo Community College** (Tsaile, AZ)
University of Arizona (Tucson, AZ)

Arkansas
University of Arkansas, Fayetteville (Fayetteville, AR)
University of Arkansas at Pine Bluff* (Pine Bluff, AR)

California
D-Q University** (Davis, CA)
University of California System (Oakland, CA)

Colorado
Colorado State University (Fort Collins, CO)

Connecticut
Connecticut Agricultural Experiment Station (New Haven, CT)
University of Connecticut (Storrs, CT)

Delaware
Delaware State University* (Dover, DE)
University of Delaware (Newark, DE)

District of Columbia
University of the District of Columbia (Washington, DC)

Florida
Florida A&M University* (Tallahassee, FL)
University of Florida (Gainesville, FL)

Georgia
Fort Valley State College* (Fort Valley, GA)
University of Georgia (Athens, GA)

Guam
University of Guam (Mangilao, GU)

Hawaii
University of Hawaii (Honolulu, HI)

Idaho
University of Idaho (Moscow, ID)

Illinois
University of Illinois (Urbana, IL)

Indiana
Purdue University (West Lafayette, IN)

Iowa
Iowa State University (Ames, IA)

Kansas
Haskell Indian Nations University** (Lawrence, KS)
Kansas State University (Manhattan, KS)

Kentucky
Kentucky State University* (Frankfort, KY)
University of Kentucky (Lexington, KY)

Louisiana
Louisiana State University System (Baton Rouge, LA)
Southern University System* (Baton Rouge, LA)

Maine
University of Maine (Orono, ME)

---

*indicates 1890 land-grant institution
**indicates 1994 tribal college land-grant institution

Maryland
University of Maryland at College Park (College Park, MD)
University of Maryland Eastern Shore* (Princess Anne, MD)

Massachusetts
Massachusetts Institute of Technology (Cambridge, MA)
University of Massachusetts (Amherst, MA)

Michigan
Bay Mills Community College** (Brimley, MI)
Michigan State University (East Lansing, MI)

Micronesia
Community College of Micronesia (Kolonia, Pohnpei, FM)

Minnesota
Fond Du Lac Community College** (Cloquet, MN)
Leech Lake Tribal College** (Cass Lake, MN)
University of Minnesota (Minneapolis, MN)

Mississippi
Alcorn State University* (Lorman MS) Mississippi State University (Mississippi State, MS)

Missouri
Lincoln University* (Jefferson City, MO)
University of Missouri System (Columbia, MO)

Montana
Blackfeet Community College** (Browning, MT)
Dull Knife Community College** (Lame Deer, MT)
Fort Belnap Community College** (Harlem, MT)
Fort Peck Community College** (Poplar, MT)
Little Big Horn College** (Crow Agency, MT)
Montana State University (Bozeman, MT)
Salish Kootenai College** (Pablo, MT)
Stone Child College** (Box Elder, MT)

Nebraska
Nebraska Indian Community College** (Winnebago, NE)
University of Nebraska System (Lincoln, NE)

Nevada
University of Nevada, Reno (Reno, NV)

New Hampshire
University of New Hampshire (Durham, NH)

New Jersey
Rutgers, The State University of New Jersey (New Brunswick, NJ)

New Mexico
Crownpoint Institute of Technology** (Crownpoint, NM)
Institute of American Indian Arts** (Santa Fe, NM)
New Mexico State University (Las Cruces, NM)
Southwest Indian Polytechnic Institute** (Albuquerque, NM)

New York
Cornell University (Ithaca, NY)

North Carolina
North Carolina A&T State University* (Greensboro, NC)
North Carolina State University (Raleigh, NC)

North Dakota
Fort Berthold Community College** (New Town, ND)
Little Hoop Community College** (Fort Totten, ND)
North Dakota State University (Fargo, NK)
Standing Rock College** (Fort Yates, ND)
Turtle Mountain Community College** (Belcourt, ND)
United Tribes Technical College** (Bismarck, ND)

Northern Marianas
Northern Marianas College (Saipan, CM)

Ohio
The Ohio State University (Columbus, OH)

Oklahoma
Langston University* (Langston, OK)
Oklahoma State University (Stillwater, OK)

Oregon
Oregon State University (Corvallis, OR)

Pennsylvania
Pennsylvania State University (University Park, PA)

Puerto Rico
University of Puerto Rico (San Juan, PR)

Rhode Island
University of Rhode Island (Kingston, RI)

South Carolina
Clemson University (Clemson, SC)
South Carolina State University* (Orangeburg, SC)

South Dakota
Cheyenne River Community College** (Eagle Butte, SD)
Oglala Lakota College** (Kyle, SD)
Sinte Gleska University** (Rosebud, SD)
Sisseton Wahpeton Community College** (Sisseton, SD)
South Dakota State University (Brookings, SD)

Tennessee
Tennessee State University* (Nashville, TN)
University of Tennessee (Knoxville, TN)

Texas
Prairie View A&M University* (Prairie View, TX)
Texas A&M University (College Station, TX)

Utah
Utah State University (Logan, UT)

Vermont
University of Vermont (Burlington, VT)

Virgin Islands
University of the Virgin Islands (St. Thomas, VI)

Virginia
Virginia Polytechnic Institute & State University (Blacksburg, VA)
Virginia State University* (Petersburg, VA)

Washington
Northwest Indian College** (Bellingham, WA)
Washington State University (Pulman, WA)

West Virginia
West Virginia University (Morgantown, WV)

Wisconsin
College of the Menominee Nation** (Keshena, WI)
Lac Courte Oreilles Ojibwa Community College** (Hayward, WI)
University of Wisconsin-Madison (Madison, WI)

Wyoming
University of Wyoming (Laramie, WY)

Please Note: of these 105 land-grant institutions, all but three (the Community College of American Samoa, the Community College of Micronesia, and Northern Marianas College) are members of the National Association of State Universities and Land-Grant Colleges. The 29 tribal colleges of 1994 are represented as a system by the single membership of the American Indian Higher Education Consortium.

Appendix 6:

U.S. Department of Agriculture Headquarters Organization

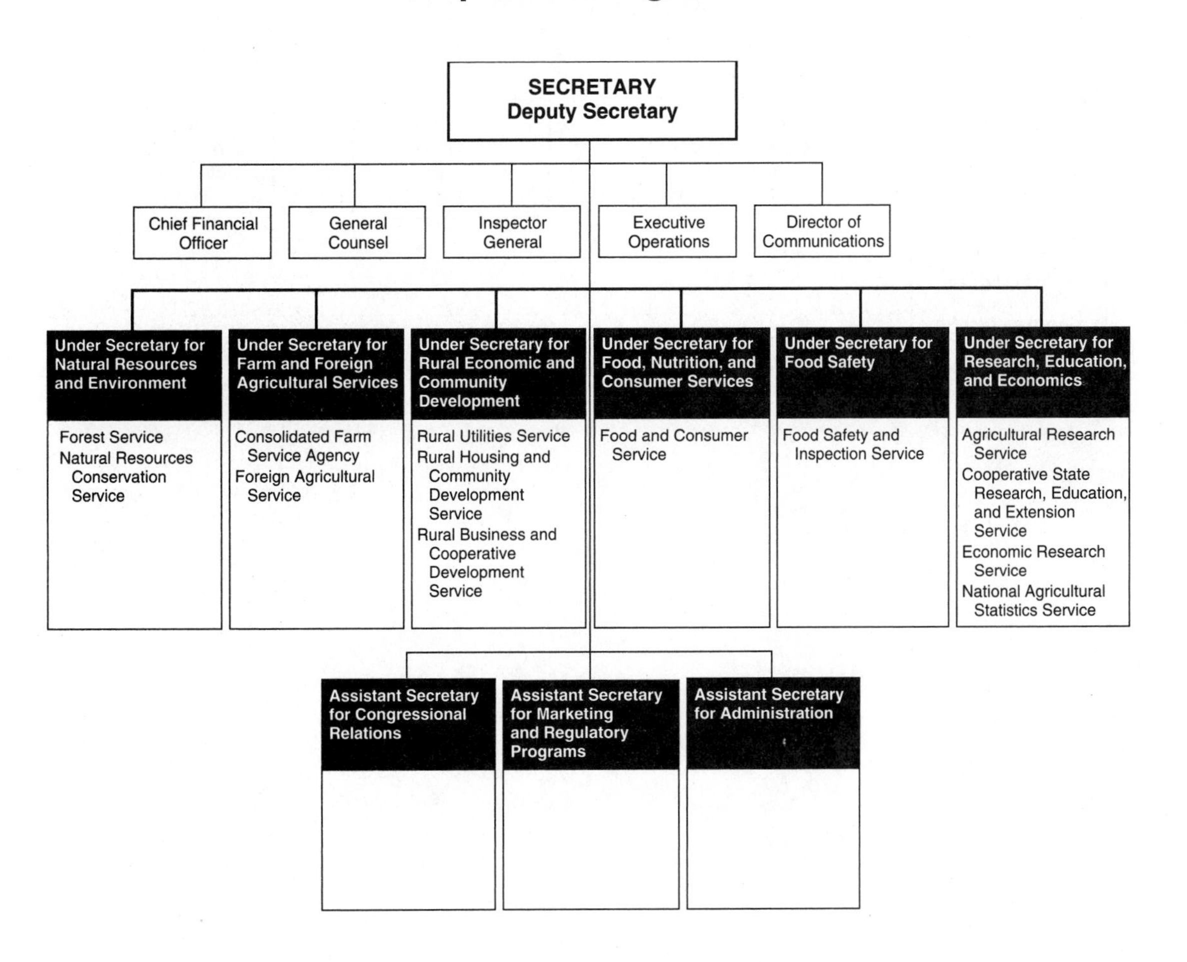

SECRETARY
Deputy Secretary
Chief Financial Officer
General Counsel
Inspector General
Executive Operations
Director of Communications
Under Secretary for Natural Resources and Environment
Forest Service
Natural Resources Conservation Service
Under Secretary for Farm and Foreign Agricultural Services
Consolidated Farm Service Agency
Foreign Agricultural Service
Under Secretary for Rural Economic and Community Development
Rural Utilities Service
Rural Housing and Community Development Service
Rural Business and Cooperative Development Service
Under Secretary for Food, Nutrition, and Consumer Services
Food and Consumer Service
Under Secretary for Food Safety
Food Safety and Inspection Service
Under Secretary for Research, Education, and Economics
Agricultural Research Service
Cooperative State Research, Education, and Extension Service
Economic Research Service
National Agricultural Statistics Service
Assistant Secretary for Congressional Relations
Assistant Secretary for Marketing and Regulatory Programs
Assistant Secretary for Administration

Research, Education, and Economics

Under Secretary

Office of the Under Secretary

Deputy Under Secretary

Agricultural Research Service (ARS)

Cooperative State Research Education, and Extension Service (CSREES)

Economic Research Service (ERS)

National Agricultural Statistic Service (NASS)

Administrative and Financial Management (AFM)

U.S. Department of Agriculture

Index

A

Academy for Educational Development (AED), 209
Accountability mandates, 168
Accuracy of an evaluation, 170
Activity-oriented individuals, 135
Administration of programs, 61–62
Administrators, 51
Adoption-diffusion process, 126–30
  adopter categories, 128–29
  adoption, defined, 126
  innovation
    characteristics of, 127–28
    defined, 126
  model, 87
  roles that influence adoption decisions, 129–30
Adult education, 123–24
  barriers to participation in, 135
Advisory Service, 219
Affective domain, 131
Affirmative action programs, 60
Agnew, Ella G., 33, 34
Agricultural demonstrations, 29–32
Agricultural educator, 73
Agricultural experiment stations, 23–24
Agricultural Extension Service, 3
Agricultural program areas in CES, 70–75
  audiences, 71–73
  components and scope, 73
  educational methods, 74
  effect on environment, concerns about, 70
  goal, 70–71
  input/output industries, 70, 71
  major programs, 74
  natural resource programs, 74–75
  share of budget, 70
Agricultural Stabilization and Commodity Service (ASCS), 72
Agriculture
  base programs in, 95
  history of, in the U.S., 15–18
  land grants, 17–18
  mission of, 9
  societies, 16–17
Agriculture of the United States, 15
Agriculturist, 73, 74
Agronomy specialists, 73
American Academy of Arts and Sciences, 16
American Agriculturist, 17
American Association of Agriculture Colleges and Experiment Stations (AAACES), 35
American Farmer, 17
American Philosophical Society, 16
American Poland-China Record Book, 28–29
Americans with Disabilities Act (ADA), 60
Analyst, 94
Andragogical model of education, 123
Appropriateness of an evaluation, 170
Assessment
  needs assessment, 98–101, 178
  of performance, 58–59, 229–32
Association for International Agricultural and Extension Education (AIAEE), 57
Atwater, Wilbur, 22

B

Bailey, Liberty Hyde, 32, 35
Bangladesh, NGOs in, 209, 219
Bankhead-Flannegan Act of 1945, 38
Base programs, 95, 116
Behavioral leadership, 65
Behaviorist educational approach, 6
Bennett's Hierarchy model, 104–6, 131
  for evaluation of programs, 172
Benson, O.H., 34
Bentley, W.D., 32
Berkshire Agriculture Society, 16
"Bill of Rights for the Adult Learner," 118
Black families, early extension efforts for, 27
Black land-grant institutions, 24, 236
Black populations, volunteer programs among, 199
Bloom's taxonomy levels of cognitive learning, 131
"Bottom-up" approach, 207
Boys' and girls' clubs. *See* 4-H/youth program
Brainstorming, 179–80
Buchanan, President, 20
Bureau of Plant Industry, 29
Butterfield, Kenyon, 35

C

Campbell, Thomas, 31
Camps, 150–51
Canada, extension in, 216
Canning clubs, 32
Card sort, 179
Carver, George Washington, 27, 31
Causal analysis, 100
Cellular radio technology, 159
Chautauqua system, 25
Checklists, 113–14
Chinese model of extension, 213–15
Civil Rights Act of 1964, Title VII, 59, 60
Clientele needs, 232
Clients-at-risk, 75–76
Clinics, 151
Club programs
  family community education, 153
  4-H, 152–53
  leader training meetings, 153–54
Cluster Action Plan, 86, 87
Clusters, 49
Coalition of Adult Education Organizations (CAEO), 118
Coalitions, 111–13
Cognitive domain, 131
Cognitive skills, 132
Colleges for the common people, 19–25
  Hatch Act of 1887, 22–24
  Morrill Act of 1862, 19–22
  Morrill Act of 1890, 24–25
Columbia University, 19
Committees
  functions of, 110
  and program development, 109–11
  purposes of, 111
Communication model, 125–32
Communities in Economic Transition, 95
Community clubs, 81
Community development, 85–87
  audiences, 85
  components of the program, 85–86
  mission and goals, 10, 85
  processes, 86–87
Community development programs, 81
Community Service Act of 1990, 199
Community services, 86
Compatibility, 127
Competency areas, CES professionals, 53, 55
Complexity, 127
Computer
  networks, 57–58
  programs, 74
Computer-Assisted Learning, 82
Computers, as teaching method, 156–57, 159
Concept mapping, 178–79
Consideration orientation, 65
Continuing Education Units (CEUs), 50, 88
Cooperative, defined, 3
Cooperative Extension System (CES)
  classification of employees, 2
  conceptual models, 10–11
  mission of, 7–10, 233
  origin of the name, 3
  overview of, 1–3
  philosophy of, 5–7, 95
  profile, 3–5
  the program today, 39–40
  source of funding for, 4
  U.S. Department of Agriculture (USDA), 1
  values and beliefs, 6–7
  during WW I & II, 38
Cooperative State Research, Education, and Extension Services (CSREES), 3, 9, 46, 95
Corn-growing contests, 33
County Agent, The, 39
County agents, 39, 40, 49, 52, 73
"County director," 51
Cromer, Marie, 34
Cross-hatch analysis, 182
Curriculum
  development, future focus of, 234
  options, 122
  See also Program Development
Czech Republic, extension in, 216–17

D

Data analysis, 182–84
Data collection methods, 175–76
Delivery methods
  future focus of, 232, 235–36
Delphi technique, 180–81
Demographics, 99
Department of Agriculture, creation of, 16
Developing countries, extension in, 207, 209–13
  constraints in, 217–19
Developmental programs, 98
Dewey, John, 5
Directors, 51
Discipline-based programming, 115–16

Discrimination, classification of, 59–60
Disparate impact, 60
Disparate treatment, 59–60
Domains of Learning, 131
Donor agencies, 208–9

E

Early adopters, 128
Early majority members, 128
Eastern Europe, extension in, 216–17
Economic development programs, 86
Economic factors, and program planning, 96–97
Economic restructuring, 227
Economic specialists, 73
ECOP, 46
Education
  agricultural demonstrations, 29–32
  boys' and girls' clubs (4-H), 32–34
  Chautauqua system, 25
  demonstration work with women, 34–35
  extension changing, but unchanging, 37–40
  Farmer's Institutes and "moveable schools," 25–26
  idea of "university extension" courses, 25
  internationalizing extension programs, 204–5
  Lyceum movement in, 25
  Seaman A. Knapp, 28–29
  Smith-Lever Act, 35–40
  See also Staff development; Teaching-learning process; Teaching methods; Training
Educational aids to teaching, 159–60
Educational approaches, 6
Educational factors, and program planning, 96–97
Effectiveness, 173
Efficiency, 173
Electronic aids to teaching, 160–61
Ellsworth, Henry L., 16
Emotional factors
  and program planning, 96–97
  that influence learning, 133
Employment in CES, 53–54, 55
Encourager, 94
Energy programming, 87–88
Engineering specialists, 73
Entomology specialists, 73
Environment, agricultural effects on, 70
Epsilon Sigma Phi, 6, 7, 57
Equal Employment Opportunity Commission (EEOC), 59
Esteem, 135
Ethical issues, in program planning, 116–18
Evaluating extension programs, 165–84
  accountability mandates, 168
  accuracy of an evaluation, 170
  brainstorming, 179–80
  card sort, 179
  data analysis, 182–84
  Delphi technique, 180–81
  evaluation, defined, 165–66
  evaluation components, 166
  evaluation standards, 169–70
  generic evaluation format, 174
  group techniques
    concept mapping, 178–79
    focus groups, 177–78
    nominal group process, 176–77
  managing an evaluation, 168–70
  objective-based evaluations, 171–75
    collecting evidence, 173–75
    creating indicators, 173
    data collection methods, 175–76
    hierarchy of evidence, 172
    questionnaires, constructing, 175–76
    role of objectives, 171
  performance measurement model, 168, 169
  preliminary questions, 166
  program monitoring, 181
  program reviews, 180
  qualitative data, 174, 175
  quantitative data, 174, 175
  reasons not to evaluate, 168
  reasons to use both formative and summative evaluation, 167–68
  reasons to use formative evaluation, 167
  reasons to use summative evaluation, 167
  reporting results, 184
  self-monitoring, 181
  stakeholders role in evaluation, 169, 171
  steps in, 170–71
Evans, J.A., 32
Exhibits, 157
Expanded Food and Nutrition Educational Program (EFNEP), 37, 78, 81
Expectations of clientele, 173
Experimental farms, 22–23
Extension, defined, 3
Extension Committee on Organization and Policy (ECOP), 1, 46
Extension Committee on Policy Program Development Task Force, 118

Extension in the 80's, 8
Extension Professional's Creed, 6, 7

F
Facilitator, 94
Family and home redefined, 227
Family community education, 153
Family resource management, 77
Farm and Home Development program, 38
Farmers Cooperative Demonstration Work, 31, 36
Farmers Home Administration (FmHA), 72
Farmer's Institutes, 25–26, 33–34
Farmer's Journal, 28, 29
Farming Systems Research and Extension (FSR/E), 211–12
Farm Seed and Loan program, 38
Farm Service Agency (FSA), 72
Farm visits, teaching method, 144
Fault-tree analysis, 100
Feasibility of an evaluation, 170
Federal support for land-grant colleges, 24
Field, Jesse, 34
Field agents, 235
  first black appointed, 31
Field days, 150
Focus groups, 177–78
Focus on Integrating Newcomers into Education (FINE), 82
Food and Agriculture Act (1977), 168
Food and Agriculture Organization (FAO), 208
Food Safety and Quality, 95
Formative evaluation, 167
"Formula funds," 46
4-H/youth program, 32–34, 36, 39, 40, 75, 78–85, 151, 152–53
  activities beyond the local level, 83–84
  awards and recognition, 46, 83, 84
  education delivery methods, 78
  International Four-H Youth Exchange (IFYE), 84, 204
  mission and goals, 9–10, 78–79
  motto, 133
  national enrollments, 80
  organizational structure, 80–81
  Professional Research and Knowledge base (PRK), 84–85
  programs for younger children, 82–83
  projects, 83
  symbol and pledge, 79, 80
  target audiences, 81, 82
  youth-at-risk, 81–82
Framing the Future (1995), 8
France, extension in, 216
Franklin, Benjamin, 16
Funding, 4, 46–47, 232
  future focus of, 235
  "soft money," 46, 78
Future focus of CES
  1890 and 1994 institutions, 236
  challenges to shape future directions, 238
  clientele satisfaction, 228–29
  delivery methods, 235–36
  extension in a complex world, 225–26
  forces reshaping America, 227
  funding, 235
  internal assessments, 229–32
  the mission, 233
  national trends in Extension, 230–32
  organization and structure, 234–35
  program content, 233–34
  program planning and evaluation, 233
  societal trends, 226–28
  staffing, 237
  strategic framework, 237–38
Future Shock, 226
Futures wheel, 178–79
Futuring methods of needs assessment, 100

G
Gardiner Lyceum, Maine, 19
Generic evaluation format, 174
Globalization, 227
Goal-oriented individuals, 135
Government Performances and Results Act of 1993, 168, 228
Graham, Albert B., 32
Green, E.H.R., 30
"Green Revolution," 212
Group forum meetings, 151
Group processes, 99–100
Guidance team, 94

H
Hampton Institute, 31
Harvard College, 19
Hatch Act of 1887, 22–24
Health, personal and environmental, 227
Historical factors, and program planning, 96–97
Holden, Perry, 27

Holmes, Mattie, 34
Home economics programs, 75–78
  audiences, 75–76
  educational delivery methods, 76–77
  goals, 75
  mission of, 9
  program scope, 77–78
Homemaker clubs, 76
Homestead Act (1862), 18
Home visits, 144
Houston, David, 36
Human development, 77
Humanist educational approach, 6
Hunt, Thomas S., 32
Hunter, Annie Peters, 34

I

Imparting knowledge model, 11
India, extension in, 212–13
Individual and societal roles, redefinition of, 227
Informational barriers to learning, 135
Informational programming, 98
Information-based economy, 227
Information delivery, 142
Information-giving techniques, 143
Innovation
  characteristics of, 127–28
  defined, 126
Innovators, 128
Input/output industries, 70, 71
In-service training, 57
Institutional barriers to learning, 135
Institutional programming, 98
Integrated Model, 219
Intellectual factors, that influence learning, 133
"Interactive classroom," 159
Interdisciplinary programs, 87
International extension
  Chinese model, 213–15
  comparison of U.S. and other countries, 206–8
  constraints in industrialized countries, 219–21
  in developing countries, 207, 209–13
    constraints in, 217–19
  donor agencies, 208–9
  in Eastern bloc and former USSR countries, 216–17
  in education programs, 204–5
  Farming Systems Research and Extension (FSR/E), 211–12
  gaining ideas from other systems, 204
  in India, 212–13
  opportunities for careers and experiences abroad, 204
  origins of, 205–6
  participatory model, 212
  percentage of population in production agriculture, 208
  project approach, 211
  Training and Visit System (T&V), 210, 211
  in western Europe and Canada, 215–16
International Four-H Youth Exchange (IFYE), 84, 204
International Fund for Agriculture Development (IFAD), 208–9
Interpersonal skills, 132
Iowa State Improved Stock Breeders Association, 28
ISOTURE model, 190, 191
Issue-based programming, 88, 114–16
Issues, defined, 115

J

Japan, extension in, 205–6
Jefferson, Thomas, 16
Journal of Extension, 6, 7, 57, 229

K

Kellogg Foundation, 209
Kepner Report (1946), 8
Kern, O.J., 32
Kings College, 19
Knapp, Seaman A., 6, 23, 28–29, 31, 33, 35, 133, 139
Knowledge-applying techniques, 143
Knowles, Malcolm, 123–24

L

Laggards, 128
Land Act (1800), 18
Land-grant institutions, 2, 19–25, 46
  positioning of Extension within, national trends, 230–31
  relationships among, 238
Land grants, 17–18
Late majority members, 128
Leadership and leadership development, 64–66, 86
  definition of leadership, 64–65
  leadership styles, 65
  training meetings for, 153–54

"Learn by Doing," 133
Learner-centered model of education, 122, 123
Learning-oriented individuals, 135
Legal implications
 of employment and programming, 59–60
 of volunteer programs, 200
Lever, Asbury Francis, 35
Liberal educational approach, 6
Life Cycle Model of Program Development, 92–93
Lincoln, Abraham, 20
Local resources, use of, 122
L-O-O-P model, 190–91
Low-income families, demonstrations designed for, 146–47
Lyceum movement in education, 25

M

"Mailbox Homemakers," 76
Management
 defined, 62–63
 leadership and leadership development, 64–66
 principles of, 62–64
Maslow's Hierarchy of Needs, 134–35
Mass-media teaching methods
 advantages/disadvantages of, 160
 aids to teaching, 159–60
 computers, 159
 electronic aids to teaching, 160–61
 exhibits, 157
 newsletters, 158
 news stories, 155
 publications, 157–58
 radio, 155–56
 telephone dial access, 158–59
 television, 156–57
Master Gardener Program, 72–73
Master Sheep Producer and Livestock Master, 73
Maturation of America, 227
McKinley, William, 5
Meetings, 148–52
 camps, 150–51
 clinics, 151
 club programs, 152–54
 by distance education, 154
 group forum, 151
 4-H activities, 151
 program planning, 151–52
 short course, 149–50
 special interest, 150
 tours and field days, 150
 workshop, 150
Megatrends 2000, 226, 237
Memorandum of Understanding (1916), 36, 37, 46
Mentoring programs, 58
Method demonstration, 148, 149
Michigan State University, 19
"Micro-climate" partnerships, 234
MI/LEAD, 86
Mission of CES, 7–10
 in agriculture, 9
 in community development, 10
 conceptual models, 10–11
 in 4-H, 9–10
 in home economics, 9
Morrill, Justin, 20
Morrill Act of 1862, 1, 19–22
Morrill Act of 1890, 1, 24–25
Mosaic Society, 227
Motivation skills, 132
"Moveable schools," 27
Multi-attribute analysis, 183–84
Multi-county programming, 50
Multiplier Effect, 77

N

National Agricultural Extension Leadership Workshop (1994), 227
National Association of County Agricultural Agents (NACAA), 57
National Association of Extension 4-H Agents (NAE4-HA), 57
National Association of Extension Home Economists (NAEHE), 57
National Association of State Universities and Land-Grant Colleges (NASULGC), 46
National Extension Honorary Society, 6, 7, 57
National service initiatives, 95, 115, 116
 volunteer programs, 199
Native American populations, 236
Natural Resource Conservation Service (NRCS), 72
Natural resource programs, 74–75
Needs assessment, 98–101, 178
Netherlands, extension in, 216
Newsletters, 158
News stories, 155
Nominal group process, 176–77
Nongovernmental Organizations (NGOs), 209

North Central Region Educational Materials Project, 234
Notebooks, 113
Nutrition, 77

O

Observability, 127
Office visits, teaching method, 144
Ohio company for Improving English cattle, 16–17
Ohio 4-H volunteer standards of behavior, 195
Ohio University, 25
Organic Act (1862), 20
Organization of CES, 45–50
  funding, 46–47
  future focus of, 234–35
  organizational chart, 48
  programming, 47–48
  staffing, 50–53
  structure, 49–50, 95
Organizations, professional, 57
Orientation, 56
Otwell, Will B., 32

P

Paraprofessionals, 52
Parenting education, 77
Participatory model of international extension, 212
Partners for International Education and Training (PIET), 209
Patents Office, creation of, 1790, 16
Pedagogical model of education, 123
Pennsylvania State University, 19
People and a Spirit Report (1968), 8
People's Republic of China, extension in, 214–15
Performance appraisals, 58–59
Performance measurement model, 168, 169
Personal correspondence, teaching method, 145
Personal factors, and program planning, 96–97
Personal reading, 57
Pesticide Applicator Training, 37
Philadelphia Society for Promoting Agriculture, 16
Philosophy of CES, 5–8, 95
  educational approaches, 6
  values and beliefs, 6–7
Physical factors, that influence learning, 132–33
Physiological level of need, 135
Pierce, J.B., 31
"Plan for a State University for the Industrial Classes," 19
Planning and Conducting Needs Assessments: A Practical Guide, 100
Plan of Work (POW), 95, 114
Plight of Young Children, 95
Plough Handle Grange, 28
Political influences, on program planning, 96–97
Popcorn, Faith, 227
Population, rural, 37
Porter, Walter C., 30
Porter Community Demonstration Farm, 30
Power Shift, 226
Preservice training, 55–56
Priority setting, in program planning, 101–2
Problem-solving model, 11
Proctor, W.C., 32
"Producers," 70
Professional organizations, 57
Professional Research and Knowledge base (PRK), 84–85
Program areas in CES, 69–88, 231
  agriculture, 70–75
  community development, 85–87
  Continuing Education, 88
  energy programming, 87–88
  home economics, 75–78
  4-H/Youth program, 78–85
  interdisciplinary programming, 87
  issue-based programs, 88
  Sea Grant program, 88
  traditional, 69
Program content, future changes on, 233–34
Program development, 47–48, 91–119
  coalitions, maximizing impact with, 111–13
  committees and, 109–11
  communicating program plans, 114
  design and implementation, 92, 106–8
  of educational content, 107
  ethical issues, 116–18
  evaluation, 92, 108–9, 233
  factors influencing, 94–97
  identifying program goals, 98–101
  issue-based programming, 114–16
  management strategies, 113–14
  models, 92, 93
  pitfalls, 118–19
  planning, 92, 233
    steps in, 97–106

PROGRAM DEVELOPMENT (*continued*)
priorities setting, 101–2
program, definition of, 91
program delivery methods, 107
program planning, definition of, 92
of resource materials, 107–8
roles for the programmer, 94
target audiences and capabilities, identifying, 102–4
time line for implementation, constructing, 108
writing program objectives, 104–6
Program monitoring, 181
Program reviews, 180
Program specialists, 51–52
Progressive educational approach, 6
Project, defined, 153
Psychomotor domain, 131
Publications, 157–58
Public Law 83, 36
Public policy education, 86
Pugh, Evan, 19

Q

Qualitative data, 174, 175
Quantitative data, 174, 175
Questionnaires, constructing, 175–76
Quicksaul, J.L., 32

R

Radical educational approach, 6
Radio, 155–56
Rasmussen, Wayne D., 40
Raudabaugh, J. Neil, 6
Recollections of Extension History, 32
Relationship orientation, 65
Relative advantage, 127
Reminder lists, 113
Renewable Resources, 37
Reporting results of program evaluation, 184
Report of Results (ROR), 114
Result demonstration
and method demonstration, compared, 149
teaching method, 145–47
Rockefeller, John D., 31, 33
Rockwell, Norman, 39
Roosevelt, Theodore, 35
Rothemstead station, 19
Ruffin, Edward, 17
Rural Development Act of 1972, 85
Rutgers University, 25

S

Safety needs, 135
Scope Report (1958), 8
Sea Grant program, 88
Self-monitoring programs, 181
Sexual harassment, 60
Situational barriers to learning, 135
Situational Leadership Model (Hershey/Blanchard), 65
Skill-acquiring techniques, 143
Skinner, John Stuart, 17
Smith, Hoke, 35
Smith-Lever Act, 5, 7, 35–40, 46
accountability mandates, 168
and establishment of CES, 1
goal of, 203
Social activism, rebirth of, 227
Social factors, and program planning, 96–97
Social indicators, 99
Social needs, 135
Society for the Advancement of Agricultural Science, 29
"Soft money" funding, 46, 78
Soil Conservation Service (NCS), 72
South Carolina Society for Promoting and Improving Agriculture and Other Rural Concerns, 16
Special interest meetings, 150
Specialists, 51–52
Spoken methods of teaching, 142
Staff development, 56–58
Staffing, 50–53
Stakeholders, 92–93
role of in evaluation, 169, 171
Stallings, W.C., 31
Stimulator, 94
Stress management, 78
Structure of CES, 49–50
Sugarcane Extension Service model, 219
Summative evaluation, 167
Supervision, 63–64
Support staff, 52–53
Survey methods, 99
Sustainable agriculture, 74
System, defined, 3

T

Taiwan, extension in, 215
Task analysis worksheet, 113
Teacher-centered model of education, 123

Teachers of Agriculture, 29
Teaching-learning process, 121–36
  adoption-diffusion model, 126–30
  adult education, 123–24
  characteristics of excellent teachers, 131–32
  education, defined, 121
  effective teaching, 130–32
  formal *versus* nonformal education, 122–23
  interaction, 124–32
  learning process, 132–34
  motivation for participation, 134–35
  planning for success, 136
  teaching as communication, 125–26
Teaching methods, 139–63
  advantages/disadvantages of individual methods, 147
  classified by form of communication, 142
  classified by function, 142–43
  classified by nature of contact, 141–42
  group methods, 148–54
    advantages/disadvantages of, 154
    general meetings, 148–52
    meetings by distance education, 154
    the method demonstration, 148, 149
    organized club programs, 152–54
  individual contact, 143–47
    farm or home visit, 144
    office visits, 144
    personal correspondence, 145
    result demonstration, 145–47
    telephone calls, 145
  information-giving techniques, 143
  knowledge-applying techniques, 143
  mass-media, 155–61
    advantages/disadvantages of, 160
    aids to teaching, 159–60
    computers, 159
    electronic aids to teaching, 160–61
    exhibits, 157
    newsletters, 158
    news stories, 155
    publications, 157–58
    radio, 155–56
    telephone dial access, 158–59
    television, 156–57
  relationship of technologies to functions of Extension work, 162
  selection of, 161–63
  skill-acquiring techniques, 143
  useful and practical instruction, 139–41
Technology development model, 219, 220
Technology-transfer model, 10–11
Telephone calls, teaching method, 145
Telephone dial access, 158–59
Television, 156–57
Timeliness and usefulness of programs, 122
Toffler, Eric, 226
Tomato clubs, 34
"Top-down" approach, 207
Total Quality Management (TQM), 66, 181
Tours, 150
Trade specialists, 73
Training, 54–58
  computer networks, 57–58
  in-service, 57
  mentoring programs, 58
  orientation, 56
  personal reading, 57
  preservice, 55–56
  professional organizations, 57
  staff development, 56–58
  volunteer programs, 197
Training and Visit System (T&V), 210, 211, 219
Trait theory, 64
Treaty of Paris (1783), 15
Trialability, 127
Tribal community colleges, 236
Turner, Jonathan, 19
Tuskegee Institute, 2, 24, 31

U

United Kingdom, extension in, 216
United Way Strategic Institute, forces reshaping America, 227
University Extension system, 50
University Outreach, 238
Urban programs, 81
U.S. Agency for International Development (USAID), 209
U.S. Department of Agriculture (USDA)
  farm-related agencies, 72
  guidelines for programs, 47
  nondiscrimination statement of, 59
  organization of, 45–46
U.S. States Agricultural Society, 17
USDA. *See* U.S. Department of Agriculture (USDA)
Useful evaluation, 170

V
Video teleconferencing, 154
Visual methods of teaching, 142
Volunteer programs, 53, 187–201
  among black populations, 199
  diversity, 199
  dollar value of volunteer time, 188–89
  evaluation of performance, 198
  identification of volunteer opportunities, 192
  ISOTURE model, 190, 191
  liability and legal implications, 200
  L-O-O-P model, 190–91
  national service, 199
  Ohio 4-H volunteer standards of behavior, 195
  orientation, 195–97
  percent of Americans who volunteer, 187
  programs and roles in CES, 196
  reasons for volunteering, 189–90
  recognition of volunteer contributions, 198
  sample volunteer agreement, 193
  selection of the volunteer, 192, 194–95
  sources of potential volunteers, 194
  supervision of staff, 197–98
  training, 197
  trends in volunteering, 188
  utilization of skills and knowledge, 197

W
Washington, Booker T., 31
Washington, George, 16
Water Quality, 95
Wesleyan University, 22
Western Europe, extension in, 215–16
Western Stock Journal, 29
Wilson, James, 5, 6
Wilson, R.S., 32
Women
  demonstration work with, 34–35
  in leadership roles, 237
Worksheets, 113–14
Workshops, 150
World Bank, 209
Written methods of teaching, 142

Y
Yearbook of Agriculture, 20
Youth-at-risk programs, 81–82